QUEST FOR ZERO-POINT ENERGY

Moray B. King

Author of Tapping the Zero-Point Energy

Engineering principles
for "free energy" inventions

Quest for
Zero-Point Energy

Moray B. King

Author of *Tapping the Zero-Point Energy*

Quest For Zero-Point Energy
by Moray B. King

ISBN: 0-932813-94-1

Published by

Adventures Unlimited Press
One Adventures Place
Kempton, Illinois 60946

Printed in the United States of America

Published in association with
Paraclete Publishing
P.O. Box 859
Provo, UT 84603

www.adventuresunlimitedpress.com
www.adventuresunlimited.nl

Dedicated to all those who are stepping up at this time to uplift the human race and our planet.

Quest for Zero-Point Energy

TABLE OF CONTENTS

Quest for Zero-Point Energy

Introduction

As I read the headlines about electric power shortages in California, I recall that it was during the oil embargo of 1974 when I first discovered the concept of zero-point energy in the final chapters of Misner, Thorne and Wheeler's book, <u>Gravitation</u>. Here was discussed Wheeler's theory of geometrodynamics, where the fabric of empty space was described as a "quantum foam" of fluctuating electric fields at extraordinary energy density on the order of 10^{94} g/cm^3. All the elementary particles were considered as subtle coherence in the form of bubbles or vortices in this energy. Even the 10^{94} g/cm^3 was considered small, being a general relativistic "cutoff" of an energy flux from a higher dimensional space, which appears infinite when projected onto our three dimensional space. Infinite energy from the fabric of space sounds like science fiction, yet this is what physicists were describing. As an engineer I asked, could this really be true and if so, could the energy be tapped as an energy source? I passionately researched the physics literature while creating a series of papers, and it resulted in my 1989 book, *Tapping the Zero-Point Energy*.

The quest then turned to the question, how do we build a practical self-running device? Over the past ten years, I studied many intriguing inventions and the plight of the inventors. Many inventions often had characteristics that matched descriptions of plasma behavior in the scientific literature, which were known to exhibit energy anomalies. Abrupt high voltage discharges, abruptly bucking electromagnetic fields and the abrupt cracking of solids could sometimes create micron size plasma forms akin to ball lightning, which seemed to contain excess energy. An inventor would typically notice a small energy anomaly and then would persistently alter his apparatus, sometimes over years, to increase the output. Knowing that standard academic science could not explain his results, the inventor would often create his own theory, typically in the form of an ether theory or a novel mass to energy transformation. Ironically, many of their ideas were similar to the various theories of the vacuum energy as published in the standard literature.

Many researchers would prefer to call the underlying, energetic fabric of space "the ether." The scientific community has rejected simplistic static ether models, but Wheeler's quantum foam description of the vacuum energy (zero-point energy) can be viewed as active, dynamic ether. It has the advantage of many supporting references in the standard physics journals, and the references lead to many points of view as to the nature of the physical vacuum. By combining the study of the details of promising inventions with the theories of the zero-point energy, I noticed a pattern, which seems to yield engineering principles that could guide the development of many types of novel zero-point energy inventions. In my presentations at energy conferences and in my technical papers, I have emphasized these engineering principles.

This book is a collection of my technical papers over the past ten years. Each introduces the zero-point energy concepts and typically features an invention that I considered promising at the time. The 1991 and 1993 papers feature the electrostatic field-chopping device of William Hyde, which he claimed produced 20 kilowatts output while free running. The 1994 and 1999 papers feature Floyd Sweet's magnetic device, which was well witnessed, and claimed to produce 500 watts while free-running. It is probably the simplest energy invention in the history of the field. The 1996 paper features the plasma tubes of Paulo and Alexandra Correa. They appear to have rediscovered the operating principle behind the plasma tubes of T. Henry Moray, whose fifty-kilowatt energy device was perhaps the most famous in the field's history. The 1997 paper features Ken Shoulders' discovery of the "electrum validum," a micron size, charge plasma form that seems to contain excess energy. It might be at the heart of many energy inventions. The 1998 paper overviews the vacuum energy ideas popular in Russia. It emphasizes the importance of counter-rotation to activate large effects. The 2000 paper speculates that the vacuum energy could be organized into vortex forms at various size scales, and suggests that devices producing these forms might manifest large energetic effects. At this time I still feel that the featured inventions are worthy of further investigation.

My intention for writing this book was to show that the concept of tapping the vacuum energy could be scientifically supported with today's physics. I hope that it inspires scientists, engineers and inventors worldwide to join the quest to help discover a fantastic new source of energy.

Moray B. King (January 2001)

Acknowledgements

Many have helped by sharing their research contained in this book, and they are referenced and acknowledged in the individual papers. I would like to give a special thanks to David Faust, Neil Boyd, and Jack Slovak for many years of consulting, inspiration and support. I would also like to thank Jeff Norris whose recent support allowed the book to be completed at this time. Finally I want to thank my wife, Suzanne, whose love and devotion provided a bedrock for years of research.

Quest for Zero-Point Energy

TAPPING THE ZERO-POINT ENERGY AS AN ENERGY SOURCE

May 1991

Abstract

The hypothesis for tapping the zero-point energy (ZPE) arises by combining the theories of the ZPE with the theories of system self-organization. The vacuum polarization of atomic nuclei might allow their synchronous motion to activate a ZPE coherence. Experimentally observed plasma ion-acoustic anomalies as well as inventions utilizing cycloid ion motions may offer supporting evidence. The suggested experiment of rapidly circulating a charged plasma in a vortex ring might induce a sufficient zero-point energy interaction to manifest a gravitational anomaly. An invention utilizing abrupt E field rotation to create virtual charge exhibits excessive energy output.

Introduction

Today's physics might allow the possibility of tapping virtually limitless quantities of energy directly from the fabric of space. Such a surprising conjecture arises by merging two separate theoretical areas of modern physics: 1) The theories of the zero-point energy [1-5] (ZPE) that model the vacuum as containing real, energetic fluctuations of electric field energy, and 2) the theories of system self-organization [6-13] which not only open the possibility of inducing coherence in this energy, but also provide the underlying principles on how this could be achieved [10]. At first this hypothesis might seem to be a blatant violation of the conservation of energy. But the key question is does the zero-point energy *really* exist? If so, a real energy is already present and its conservation would not be an issue.

The real issue centers on how random fluctuations could become coherent. Any spontaneous coherence seems to violate the second law of thermodynamics, which is generally understood to mean systems should evolve toward random behavior, not toward coherence. This point is thoroughly discussed in the theories of system self-organization [11,12]. Prigogine [13] won the 1977 Nobel prize in chemistry for defining the conditions under which a system could evolve from randomness toward coherence. The conditions are that the system must be 1) far from equilibrium, 2) nonlinear in its dynamics and 3) have an energy flux through it. These conditions are expressed in general system theory terms, and it turns out that the already published theories of the ZPE can, under certain circumstances, fulfill these conditions.

Despite the intriguing possibility offered by system theory, no purely theoretical discussion could ever prove that the zero-point energy could be tapped as an energy source. Only an experiment coupled with the theory would be convincing. This article discusses how observed anomalies associated with the ion-acoustic oscillations in plasmas could be a manifestation of a coherent ZPE interaction and in particular, how the cycloid motion of a plasma's nuclei might induce a sufficient ZPE coherence to manifest a gravitational anomaly. This article also highlights an invention which utilizes the abrupt rotation of electric fields to cause an hypothesized pair production of virtual charges from the vacuum energy across a macroscopic system. The invention reportedly outputs excessive power while free running, and its full disclosure may constitute an experiment which could be repeated by the scientific community.

The Fabric of Space

Does the fabric of empty space really contain a plenum of energy? This question has been debated throughout the history of science. The early scientists through the 19th century believed in the existence of an ether, which was modeled as a material substance that could support the wave propagation of light. The famous Michelson-Morley experiment failed to detect the expected ether wind produced by the earth's motion though it. At the turn of the century Einstein used this result to support the theory of special relativity. When this became accepted, the scientific community rejected the existence of the ether. Thus classical physicists came to consider the vacuum of space to be truly empty.

The classical model was only to last until the 1930's when quantum mechanics became accepted. From quantum mechanics arose a mathematical term in the description of the ground state of any oscillating system called the zero-point energy. The term "zero-point" refers to zero degrees Kelvin which means this energy exists even in the absence of all heat. The energy was interpreted as being inherent to the fabric of space itself. Dirac [14] showed how electron-positron pair production could arise from the vacuum fluctuations and quantum electrodynamics was born. The Heisenberg uncertainty principle allowed quantum mechanical systems to "borrow" this energy for short periods of time. The ether came back into science not modeled as a material substance but rather as a randomly fluctuating energy.

Could a space filled with fluctuations of electric flux be consistent with special relativity? Boyer [15] showed that, by invoking the postulate of Lorentz invariance, the spectral energy density D of the zero-point fluctuations must have the particular form as a function of frequency f

$$D(f) = kf^3$$

where the constant k is related to Planck's constant. This result gives a quantitative basis to the theory of random electrodynamics which strives to show that quantum mechanical effects arise from matter's interaction with the zero-point energy.

This cubic frequency relation implies an absurd result: the energy density of the ZPE at each point in space is infinite. A similar problem plagues quantum electrodynamics where infinities are renormalized away. Some type of frequency cutoff is required to create a finite, quantitative theory. Wheeler [16] applied the theory of general relativity to the ZPE to create a natural cutoff in his theory of geometrodynamics. In general relativity the fabric of space curves as a function of energy density. When the density becomes sufficiently great, space pinches like it's forming a black hole. This gives rise to the formation of hyperspace structures that Wheeler called "wormholes." His calculation yielded microscopic channels on the order of 10^{-33}cm having a (mass equivalent) energy density of 10^{94} grams/cm^3. The resulting view is that the fabric of space consists of constantly forming and annihilating pairs of microscopic "mini" blackholes and whiteholes which channel electric flux into and out of our three dimensional space. These mini holes manifest dynamics which could be modeled as a turbulent, virtual plasma that Wheeler calls the "quantum foam." In this view the elementary particles are like bubbles or vortices arising from the dynamics of the vacuum energy.

Is it possible to tap this energy? At first the answer seems to be no since it is extremely difficult experimentally to observe its existence; the energy is ubiquitous and a detector requires an energy difference to measure field strength. However, the theories of quantum electrodynamics indicate that all the elementary particles are dynamically interacting with the ZPE resulting in vacuum polarization. In particular, quantum electrodynamics shows that the different elementary particles polarize the vacuum differently [17-19]. In a first order model, electrons, especially conduction band electrons, exhibit an ethereal cloud-like random interaction with the zero-point fluctuations and are effectively in thermodynamic equilibrium with it. No net energy would be absorbed by this type of system. However, an atomic nucleus exhibits a pattern of quasi-stable vacuum polarization channels converging toward it. This may allow the possibility of driving the nucleus-ZPE system off of equilibrium by abrupt motion. This fulfills the first condition for system self-organization.

How the other conditions could be fulfilled as well can be understood by modeling the ZPE as a virtual plasma. Like a plasma, it is nonlinear in its dynamical behavior, it may be driven off of equilibrium by the abrupt motion of nuclei, and it might well be sustained by an energy flux intersecting our three dimensional space from a higher dimensional superspace [20-22]. This last point is clearly the most speculative. If true, it offers virtually limitless energy. It can best be supported by noting that there are interpretations of quantum mechanics and relativity theory which imply the existence of a physically real, higher dimensional space, and the notion of superspace is well discussed in the physics literature [23-25]. It is interesting to note that some authors [26,27] recognized that the mathematical analysis of a nonlinear system interacting with the ZPE shows that energy could be extracted, but they are skeptical due to the lack of experimental evidence.

Ion-Acoustic Oscillations

The real proof that the zero-point energy could become an energy source can only come from a repeatable experiment. The above discussion suggests that the motion of a plasma's nuclei might be an effective transducer for interacting with the ZPE. The coherent oscillations of nuclei in a plasma is known as the ion-acoustic mode, and it has been associated with anomalous plasma behavior including run-away electrons [28], anomalous heating [29-31], anomalous resistance [32], and high frequency voltage spikes [33-35]. Could these anomalies be associated with a direct ZPE interaction manifesting a macroscopic vacuum polarization [36]? The inventor T. Henry Moray [37] stressed the importance of ion oscillations in the plasma tubes of his invention that produced 50 kilowatts of anomalous electrical power in the 1930's. His well-witnessed invention could not be explained with the physics of that time, and puzzled all the scientists who investigated his device.

Another experiment where coherent oscillations of nuclei could be the source of anomalous heat is the electrolytic "cold fusion" experiment of Pons and Fleischmann [38]. In this experiment deuterium nuclei occupy shallow potential wells in the crystal lattice sites of the palladium. Here the nuclei are free to oscillate [39], but they generally diffuse to adjacent, vacant lattice sites [40]. However, under the conditions of deuterium supersaturation all the lattice sites are occupied, and the deuterons within a crystal grain of palladium could then undergo synchronous oscillations similar to ion-acoustic oscillations. This could create a coherent ZPE interaction yielding anomalous heat [41]. This hypothesis predicts the effect would be greatly enhanced by supersaturating a pure single crystal of palladium and that an electrical pulse could trigger the oscillations. It might also be possible to generate anomalous heat with experiments using ordinary water (although it is more difficult to constrain protons to the lattice sites than deuterons). The difficulty in repeating the heat anomaly of the Pons/Fleischmann experiment is probably directly related to the well known difficulty in achieving supersaturation and failing to use pure crystalline palladium. Nonetheless this is probably the first repeatable experiment in which at least some other scientists are able to produce an energy anomaly [42].

Plasma Spirals

Other investigators have claimed energy anomalies associated with plasma behavior. The Russian plasma physicist, Chernetskii, from his observations of anomalous energetic plasma activity explains that under appropriate conditions a plasma interacts directly with the ZPE [43]. He has recently claimed to have created a plasma device that absorbs energy from the vacuum fluctuations when the plasma's particles undergo cycloid motion [43]. Likewise the inventions of Searl [44], Spence [45], and Papp [46] also have cycloid particle motion in the plasmas within their energy producing devices.

Ball lightning [47] is a possible candidate for a ZPE interaction since it has been modeled as a vortex ring plasmoid [48]. The energy source needed to maintain its persistence must be localized within the ball since it has been observed inside of shielded environments such as aircraft and submarines. In a submarine a particular type of circuit breaker has launched it on multiple occasions [49]. The vortex ring model for ball lightning has its plasma particles undergoing precessional cycloid motion, and it might therefore be an example of a zero-point energy coherence occurring in nature.

It may also be possible to induce the cycloid motion of nuclei within solid state magnetic materials such as ferrites. When a ferrite's magnetic domain wall moves, the microscopic magnetic dipoles rotate [50]. This supports the propagation of nonlinear spinor waves through the ferrite [51]. This wave directly couples to the ferrite lattice causing an elastic, acoustical spinor wave [52]. This results in the helical motion of the ferrite's nuclei. If such motion induces a zero-point energy coherence, then nearby pickup coils might detect anomalous energy. Such a hypothesis may help explain the "free energy" inventions of Coler [53] and Sweet [54].

The plasma vortex-ZPE hypothesis could also be applied to the water vortex studies of Schauberger [55]. He claimed that water forced to precess through specially shaped spiraling tubes induced an energy anomaly causing a peculiar bluish glow to appear at the center of the vortex. Also the gyroscope studies by Laithwaite [56] may fit the vortex hypothesis. Laithwaite ob-

Figure 1 Vortex ring manifesting precessional flow.

served that a precessing gyroscope that was displaced along a particular cycloid path would exhibit an inertial/gravitational anomaly.

Gravitational Anomalies

The expectation of gravitational anomalies associated with coherence of the zero-point energy arises directly from general relativity. Gravity is described as curvature of the space-time metric induced by the stress-energy tensor [57]. If the zero-point energy has the enormous density as predicted by Wheeler, then even a slight coherence in its activity could curve the local space-time metric producing measurable gravitational or time anomalies. An experiment which altered the pace of time near the apparatus would suggest the ZPE's involvement [58]. Puthoff [59] has recently quantitatively explored Sakarov's suggestion that gravity is intimately coupled to the behavior of the ZPE by proposing a model in which gravity directly arises from the action of the zero-point fluctuations. An experiment which produced a gravitational or time anomaly would yield convincing evidence that the ZPE is being cohered because the ZPE is the only energy appreciable enough to induce a space-time metric curvature by technological means.

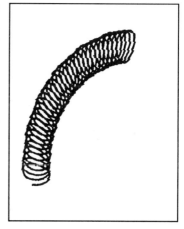

How could such an anomaly be demonstrated experimentally? The ideas presented in this paper suggest the following preliminary experiment: A piping system is shaped into a vortex ring (Figure 1) whose poloidal/toroidal size ratio is similar to the plasmoids observed by Bostick [60] in his experiments. Charged fluid or plasma is pumped to circulate rapidly through the vortex ring. Note that the plasma is forced to undergo an effective

Figure 2 Vortex filament model of abruptly rotated E field line.

precessional motion (a poloidal rotation closing into a toroidal rotation). A weight change in the apparatus or a change in the pace of time nearby the apparatus would support the proposed conjecture that an ionic plasma vortex could induce a ZPE coherence.

An oscillatory ion-acoustic plasma vortex ring can be created with an

An oscillatory ion-acoustic plasma vortex ring can be created with an electrical circuit. A toroidal coil is wound on a ferrite core with wire whose insulator is coated with a mildly radioactive material. Alternatively the coil could be bombarded by ionizing radiation [61]. The radiation only needs to be strong enough to ionize the air or gas near the surface of the toroidal coil, and it maintains a cold plasma. The coil is then tuned to resonate at the ion-acoustic frequency of this plasma by adding an appropriate capacitance to the circuit. A properly tuned resonance yields ion oscillatory displacement currents in the medium surrounding the wire which acts as a wave guide. During resonance further ionization could accrue shifting the ion-acoustic frequency. This nonlinear effect can be stabilized with a parallel, variable capacitor controlled via feedback by the magnitude of the output current. The capacitance is automatically adjusted to maximize the output current unless it becomes too large at which point the system is intentionally detuned. If the ion-acoustic plasma vortex were to interact coherently with the ZPE, then anomalous energy production might occur in such a system.

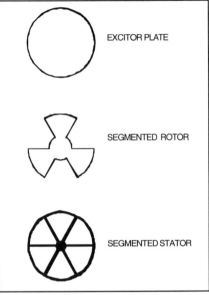

Figure 3 Simplified version of excitor, rotor and stator.

Macroscopic Pair Production

The plasma vortex ring motif can be applied directly to the virtual ZPE plasma to create a model of an elementary charge. Bostick showed that a pair of plasmoid vortex rings could arise from an abruptly excited, turbulent plasma [60]. In a similar fashion could electron-positron pair production arise as vortex rings from the ZPE modeled as a virtual plasma? In this analogy the charge would be associated with the helicity of the electric flux circulation on the vortex ring [62,63]. Likewise the electric (E) field lines emanating from a charge could be modeled as helical filaments [64]. Here the helical filaments would originate from the charge and be sustained continuously by electric flux flowing at the speed of light. This model of E field lines offers a dynamic possibility for activating the vacuum energy. If an E field line alone could be abruptly rotated, it would mimic the precessional flow of a vortex ring section (Figure 2) and consequently would manifest for an instant virtual charge at a macroscopic level. This would constitute a coherence in the zero-point energy.

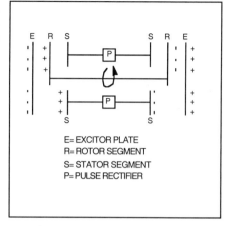

Figure 4 Connection of components (side view).

An experiment in which E field lines are abruptly rotated might yield excessive energy from the resulting voltage transients. Such an experiment has already been done, and its description is essentially the invention by Hyde [65]. Hyde uses rapidly spinning segmented rotors to abruptly cut E field lines, and his invention is reported to output power ten times the power input. The invention consists of a pair of excitor plates, a pair of segmented rotors and a pair of segmented stators (Figure 3). Charge is free to migrate on the conductive surfaces comprising the rotors and exciter plates, but on the stators the adjacent, conductive segments are electrically insulated from each other.

The components are connected as shown in the side view (Figure 4). An external voltage source charges the exciter plates which provide an electrostatic polarization field. Insulation on the device's negatively charged surfaces insures that no current leaks from the exciter plates

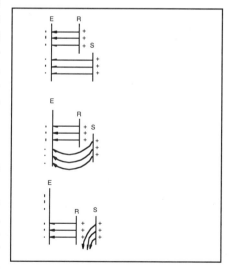

Figure 5 Abrupt E field cutting sequence (top view).

and little power is drawn from the charging voltage source. The rotors are electrically connected to each other through a conductive shaft which is spun by an electric motor (not shown). The rotors in the patent description were electrically connected via brushes to their adjacent stators, but Hyde has since improved his invention by removing these contacts [66]. The field from the exciter plates induces a polarization between the connected rotors. The segments on both rotors are aligned to allow them to shield an aligned pair of stator segments. As the rotor spins, aligned stator segments are alternately exposed to and shielded from the exciter polarization field. Each stator segment is electrically connected to its counterpart on the other stator through a pulse rectifier circuit in which the transient voltage pulses are stepped down and then channeled to a combining rectifier output circuit (not shown).

It is a surprise that such a simple device as Hyde's could output anomalous power. An analysis using just classical physics would predict that the voltage induced across a stator segment pair would swing between zero and the exciter plate voltage since this is the steady state limit for the shielded and exposed conditions. This is observed when the rotor is spun slowly. However, when the rotor is spun on the order of 6000 rpm, a 3 KV potential across the exciter plates yields stator pulses in excess of 300 KV with a very small drag on the rotor [67]. How the ZPE could be cohered by the abrupt field cutting from the rotors is illustrated in the top view sequence in Figure 5. During the exposed condition, current flows to charge the stator segment pair. Under rapid spin, the rotor blade cuts through the gap quicker than the charge can leave the stator segment due to the current's momentum from residual inductance of the connecting circuit. The charge remains on the stator segment during the instant its E field lines are cut resulting in their abrupt rotation. If such an abrupt rotation of E field lines manifests virtual charge from the vacuum energy, then this charge would greatly augment the potential across the stator segment pairs and yield a more vigorous voltage transient. Note that opposite virtual charge is created simultaneously on the outer surfaces of the connected stator segments. Quantum electrodynamics allows virtual charge pair production from the ZPE as long as charge is conserved. This analysis of Hyde's invention suggests virtual charge pair production in the macroscopic realm. The resulting transient zero-point energy coherence accelerates the charges of the stator segment circuit, and the system outputs anomalously excessive energy.

Summary

Applying the principles of system self-organization to the theories of the zero-point energy suggests that an appropriate system might be able to induce a coherence in the action of the zero-point energy. Quantum electrodynamics shows that the ZPE intimately interacts with the various elementary particles with differing vacuum polarization dynamics. The vacuum polarization description of atomic nuclei suggests that abrupt, synchronous motion of ions or nuclei may be a good candidate for coherent vacuum energy activation. The observed anomalies associated with the ion-acoustic oscillations of a plasma might be evidence for this. Further circumstantial evidence may arise from the claims of different investigators and inventors whose devices exhibit a common modus operandi: They utilize coherent, synchronous motion of ions or nuclei. The largest claims are associated with devices that produce cycloid or precessional motion of nuclei. This leads to the hypothesis that a positively charged plasma vortex might induce a ZPE coherence. The idea can be experimentally explored by rapidly circulating a charged plasma or fluid through a vortex ring piping system and looking for a gravitational or time variation since there is a recognized theoretical connection between gravity and the action of the zero-point energy.

Since vortex ring plasmoid pair production is observed in turbulent plasmas, modeling the ZPE as a turbulent, virtual plasma supports the vortex ring model for elementary charge and the vortex filament model for electric field lines. Such a model predicts that the abrupt rotation of electric field lines would manifest virtual charge from the vacuum energy. Experimental support that macroscopic, virtual charge pair production might provide energy directly from the ZPE arises from Hyde's fully disclosed invention. It appears imperative that Hyde's invention be replicated, for only a repeating experiment could prove that it is possible to tap the zero-point energy as an energy source.

Acknowledgments

The help of Oliver Nichelson, Andrea Powell and Carl Rhoades is gratefully acknowledged.

References

[1] T.H. Boyer, "Random Electrodynamics: The theory of classical electrodynamics with classical electromagnetic zero-point radiation," Phys. Rev. D 11(4), 790-808 (1975).

[2] E.M. Lifshitz, L.P. Pitaevskii, *Statistical Physics, Part 2*, Pergamon, Oxford, 1980. "Hydrodynamic Fluctuations," pp. 369-73.

[3] P.B. Burt, *Quantum Mechanics and Nonlinear Waves*, Harwood Academic, NY, 1981.

[4] L. de la Pena, A.M. Cetta, "Origin and Nature of the Statistical Properties of Quantum Mechanics," Hadronic J. Suppl. 1(2), 413-39 (1985).

[5] I.R. Senitzky, "Radiation-Reaction and Vacuum Field Effects in Heisenberg - Picture Quantum Electrodynamics," Phys. Rev. Lett. 31(15), 955 (1973).

[6] G. Nicolis, I. Prigogine, *Self Organization in Nonequilibrium Systems*, Wiley, NY, 1977.

[7] H. Haken, *Synergetics*, Spinger Verlag, NY, 1971.

[8] M. Suzuki, "Fluctuation and Formation of Macroscopic Order in Nonequilibrium Systems," Prog. Theor. Phys. Suppl. 79, 125-140 (1984).

[9] A. Hasegawa, "Self-Organization Processes in Continuous Media," Adv. Phys. 34(1), 1-42 (1985).

[10] M.B. King, *Tapping the Zero-Point Energy*, Paraclete Publishing, Box 859, Provo, UT 84603, 1989.

[11] S. Firrao, "Physical Foundations of Self-Organizing Systems Theory," Cybernetica 17(2), 107-24 (1984).

[12] Yu. L. Klimontovich, M.V. Lomonosov; "Entropy Decrease During Self-Organization and the S Theorem," Sov. Tech. Phys. Lett. 9(12), 606-7 (1983).

[13] I. Prigogine, I. Stengers, *Order Out of Chaos*, Bantam Books, NY, 1984.

[14] P.A. Dirac, Roy. Soc. Proc. 126, 360 (1930). Also G. Gamow, *Thirty Years that Shook Physics*, Doubleday, NY, 1966.

[15] T.H. Boyer, "Derivation of the blackbody radiation spectrum without quan-

tum assumptions," Phys. Rev. 182(5), 1374-83 (1969).

[16] J.A. Wheeler, *Geometrodynamics*, Academic Press, NY, 1962.

[17] F. Scheck, *Leptons, Hadrons and Nuclei*, North Holland Physics Publ., NY, 1983; pp. 212-23.

[18] W. Greiner, "Dynamical Properties of Heavy-Ion Reactions - Overview of the Field," S. Afr. J. Phys. 1(3-4), 75 (1978).

[19] J. Reinhardt, B. Muller, W. Greiner, "Quantum Electrodynamics of Strong Fields in Heavy Ion Collisions," Prog. Part. and Nucl. Phys. 4, 503 (1980).

[20] S.W. Hawking, "Wormholes in Spacetime," Phys. Rev. D 37(4), 904-910 (1988).

[21] S.W. Hawking, R. LaFlamme, "Baby Universes and the Non-Renormalizability of Gravity," Phys. Lett. B 209(1), 39-42 (1988).

[22] D.H. Freedman, "Maker of Worlds," Discover 11(7), 46-52 (July 1990).

[23] B.S. DeWitt, N. Graham, *The Many Worlds Interpretation of Quantum Mechanics*, Princeton University Press, Princeton, 1973.

[24] F.A. Wolf, *Parallel Universes*, Simon & Schuster, NY, 1988.

[25] P. Davies, *Other Worlds*, Simon & Schuster, NY, 1982.

[26] T.H. Boyer, "Equilibrium of random classical electromagnetic radiation in the presence of a nonrelativistic nonlinear electric dipole oscillator," Phys. Rev. D 13(10), 2832-45 (1976).

[27] S.I. Putterman, P.H. Roberts, "Random Waves in a Classical Nonlinear Grassman Field," Physica 131 A, 51-63 (1985).

[28] Y. Kiwamoto, H. Kuwahara, H. Tanaca, "Anomalous Resistivity of a Turbulent Plasma in a Strong Electric Field," J. Plasma Phys. 21(3), 475 (1979).

[29] J.D. Sethian, D.A. Hammer, C.B. Whaston, "Anomalous Electron-Ion Energy Transfer in a Relativistic-Electron Beam Heated Plasma," Phys. Rev. Lett. 40(7), 451 (1978).

[30] S. Robertson, A. Fisher, C.W. Robertson, "Electron Beam Heating of a Mirror Confined Plasma," Phys. Fluids 32(2), 318 (1980).

[31] M. Tanaka, Y. Kawai, "Electron Heating by Ion Acoustic Turbulence in Plasmas," J. Phys. Soc. Jpn. 47(1), 294 (1979).

[32] Y. Kawai, M. Guyot, "Observation of Anomalous Resistivity Caused by Ion Acoustic Turbulence," Phys. Rev. Lett. 39(18), 1141 (1977).

[33] Yu G. Kalinin, et al., "Observation of Plasma Noise During Turbulent Heating," Sov. Phys. Dokl. 14(11), 1074 (1970).

[34] H. Iguchi, "Initial State of Turbulent Heating of Plasmas," J. Phys. Soc. Jpn. 45(4), 1364 (1978).

[35] A. Hirose, "Fluctuation Measurements in a Toroidal Turbulent Heating Device," Phys. Can. 29(24), 14 (1973).

[36] M.B. King, "Macroscopic Vacuum Polarization," Proc. Tesla Centennial Symposium, International Tesla Society, Colorado Springs, 1984; pp. 99-107. Also reference 10, pp. 57-75.

[37] T.H. Moray, J.E. Moray, *The Sea of Energy*, Cosray Research Institute, Salt Lake City, 1978.

[38] M. Fleischmann, S. Pons, "Electrochemically Induced Nuclear Fusion of Deuterium," J. Electroanal. Chem. 261, 301 (1989).

[39] T. Springer, "Investigations of Vibrations in Metal Hydrides by Neutron Spectroscopy," in G. Alefeld, J. Volkl, *Hydrogen in Metals I, Basic Properties*, Springer Verlag, NY, 1978; pp. 75-100.

[40] R.C. Bowman, "Hydrogen Mobility at High Concentrations" in G. Bambakidis, *Metal Hydrides*, Plenum Press, NY, 1981; pp. 109-144.

[41] M.B. King, "Electrolytic Fusion: A Zero-Point Energy Coherence?" reference 10, pp. 143-166.

[42] H. Fox, "Cold Fusion Successes: Achievements and Primary Sources," Fusion Facts 1(12), Fusion Information Exchange, Salt Lake City, June 1990.

[43] A. Smokhin, "Vacuum Energy - A Breakthrough?" NOVOSTI press release 03NT0-890717CM04, 17 July 1989. Also Spec. Sci. Tech. 13(4), 273 (1990).

[44] W.P. Baumgartner, Energy Unlimited, Issue 20, Energy Unlimited, Albuquerque, NM, 1986.

[45] G.M. Spence, "Energy Conversion System," U.S. Patent No. 4,772,816 (1988).

[46] J. Papp, "Noble Gas Engine," U.S. Patent No. 4,428,193 (1984).

[47] S. Singer, *The Nature of Ball Lightning*, Plenum Press, NY, 1971.

[48] P.O. Johnson, "Ball Lightning and Self Containing Electromagnetic Fields," Am. J. Phys. 33, 119 (1965). Also R.C. Jennison, "Relativistic Phase-locked Cavity Model of Ball Lightning," Electronics Laboratory, University of Kent, U.K., 1990.

[49] P.A. Silberg, "Ball Lightning and Plasmoids," J. Geophys. Res. 67(12), 4941 (1962).

[50] T.H. O'Dell, *Ferromagnetodynamics*, John Wiley, NY, 1981.

[51] D.H. Jones, et al., "Spin-wave theory of the zero-point energy of solitons in one dimensional magnets," J. Phys. Condens. Matter 1, 6131-44 (1989).

[52] M. Cieplak, L.A. Turski, "Magnetic solitons and elastic kink-like excitations in compressible Heisenberg chain," J. Phys. C: Solid St. Phys. 13, L 777-80 (1980).

[53] "The Invention of Hans Coler, Relating to an Alleged New Source of Power," British Intelligence Objectives Sub-Committee, Final Report No. 1043, 1946. (Declassified 1962).

[54] F.S. Sweet, "Nothing is Something: The Theory and Operation of a Phase-Conjugated Vacuum Triode", June 1988; private communication, December 1989.

[55] B. Frokjaer-Jensen, "The Scandinavian Research Organization on Non-Conventional Energy and The Implosion Theory (Viktor Schauberger)," Proc. First International Symposium on Nonconventional Energy Technology, Toronto, 1981; pp 78-96.

[56] J. Davidson, *The Secret of the Creative Vacuum*, C.W. Daniel Co. Ltd., Essex, England, 1989; pp. 258-262.

[57] C.W. Misner, K.S. Thorne, J.A. Wheeler, *Gravitation*, Freeman, NY 1970. Chapters 43 and 44 discuss the zero-point energy.

[58] N.A. Kozyrev, "Possibility of Experimental Study of the Properties of Time," Joint Publication Research Service, Arlington, VA, 1968.

[59] H.E. Puthoff, "Gravity as a Zero-Point Fluctuation Force," Phys. Rev. A 39(5), 2333 (1989).

[60] W.H. Bostick, "Experimental Study of Plasmoids," Phys. Rev. 106(3), 404 (1957).

[61] P.M. Brown, "Apparatus for Direct Conversion of Radioactive Decay Energy to Electrical Energy," U.S. Patent No. 4,835,433 (1989).

[62] W.H. Bostick, "The Gravitational Stabilized Hydrodynamic Model of the El-

ementary Particle," Gravity Research Foundation, New Boston, NH, 1961.

[63] N.J. Medvedeff, *Nuclear Dynamics*, privately published, Hanover, MA, 1961.

[64] J.J. Thomson, *Recollections and Reflections*, Cambridge University Press, 1936; pp. 94, 369.

[65] W.W. Hyde, "Electrostatic Energy Field Power Generating System," U.S. Patent No. 4,897,592 (1990).

[66] W.W. Hyde, private communication, April 1991.

[67] W.W. Hyde, private communication, April 1991. A 1987 prototype containing approximately 2000 capacitors and diodes exhibited the following characteristics:

No. rotor segments	240
No. stator segments	480
Rotor speed	6000 rpm
Output voltage	602 V DC
Output current	38 amps
Output power	22.9 KW
Input power	2.4 KW
Net output power (while free running)	20.5 KW

FUNDAMENTALS OF A ZERO-POINT ENERGY TECHNOLOGY

May 1993
(Updated 1995)

Abstract

The vacuum polarization of atomic nuclei may trigger a coherence in the zero-point energy (ZPE) whenever a large number of nuclei undergo abrupt, synchronous motion. Experimental evidence arises from the energy anomalies observed in heavy-ion collisions, ion-acoustic plasma oscillations, sonoluminescence, fractoemission, large charge density plasmoids, abrupt electric discharges, and light water "cold fusion" experiments. Further evidence arises from inventions that utilize coherent ion-acoustic activity to output anomalously excessive power. A ZPE coherence sufficient to manifest a gravitational anomaly might occur from circulating charged plasma through a helical vortex ring. Abruptly pulsed, opposing electromagnetic fields may further augment any ZPE interaction.

Introduction

Modern physics views the vacuum of empty space not as a void but as a plenum of randomly fluctuating electromagnetic fields known as the zero-point energy (ZPE). These vacuum fluctuations persists at zero degrees Kelvin, and many physicists have shown they possess an enormous energy density (Hathaway, 1991). For example Wheeler (1962) derives a density of 10^{94} grams/cm^3 for individual fluctuations on the scale of a Planck length (10^{-33} cm). The energy is not readily noticed at the classical level since the fluctuations are ubiquitous, and a standard radiation detector requires an energy difference to make a measurement. Nonetheless there is a point of view well represented in the physics literature that shows the ZPE is the basis of the material world. Boyer

(1975) shows how the blackbody radiation spectrum as well as other quantum events is attributed to the ZPE. Senitsky (1973) has suggested that an elementary particle's very existence is intertwined with the ZPE. Puthoff (1987) has shown the hydrogen atom's stability is due to a ZPE interaction which prevents the electron from collapsing into the nucleus. Puthoff (1989) has also shown how gravity can be derived from the action of the ZPE, and recently Haisch, Rueda and Puthoff (1994) have shown that the ZPE could be the basis of inertia as well. There have also been proposed ZPE models for the photon (Scully, 1972, Honig, 1986) and the electron (Jennison, 1978). The zero-point energy might well be the primary underpinning of all physics.

Can the zero-point energy be tapped as an energy source? At first this idea seems to be a blatant violation of conservation of energy. But if the ZPE is physically real, then there is energy available and its conservation would not be the issue. The real issue centers on the second law of thermodynamics, the law of entropy, for how could a system based on chaotic energy fluctuations evolve into coherence? Prigogine (1977) won the Nobel prize in chemistry for showing how a system can evolve from chaos into order. The system must exhibit three characteristics: 1) It must be nonlinear, 2) far from equilibrium and 3) have an energy flux through it. A review (King, 1991) of the published theories of the zero-point energy show that under certain circumstances the ZPE in its nonlinear interaction with matter can be influenced to fulfill these conditions, and this suggests the possibility that it might be available as an energy source (King, 1989). For example Boyer (1976) has mathematically shown that a system consisting of a nonlinear dipole can absorb energy from particular modes of the ZPE spectrum. Also Cole and Puthoff (1993) published a proof showing that, in principle, the ZPE can be tapped as an energy source without violating thermodynamics. As chaos theory and system theory are applied to the ZPE, new models of matter, fields and spacetime arise (e.g. LaViolette, 1985, Winterberg, 1990). Combining the theories of system self-organization (e.g. Suzaki, 1984, et al.) with the theories of the zero-point energy open the possibility for a new energy source.

Experiments

Despite the intriguing possibilities offered by system theory, it will require an experiment to prove that the ZPE can be tapped as an energy source. A first order experimental success has already been accepted by the physics community. Forward (1984) has invented a rather simple battery based on the Casimir effect (Milonni, et al., 1988). Casimir predicted and experimentally demonstrated that the zero-point fluctuations induced a $1/d^4$ (d=distance) attraction between two parallel conductive plates. Forward's battery utilizes charged foils whose spacing is so close that the $1/d^4$ attraction overcomes the $1/d^2$ Coulomb repulsion and results in a direct current output of power from the zero-point energy. Even though the device may not currently be practical, its primary value is that it proves in principle that the ZPE can be tapped as an energy source (Puthoff, 1990).

Puthoff (1990) also applies the same Casimir squeeze principle to explain the surprising stability of high charge density, electron beads discovered by Shoulders (1991) called "electrum validum." The beads are on the order of a micron in size and exhibit a net charge of approximately 10^{11} electrons. They are stable when guided along etched channels in a dielectric substrate and can travel at one tenth the speed of light. They can induce an excessively powerful electrical pulse on a nearby serpentine conductor (or surrounding helical conductor) as they rapidly speed by in parallel to it. When they strike an anode plate, they also discharge a powerful pulse. Shoulders states that the output energy of the pulses greatly exceeds the input energy needed to create them, and the source appears to be from the zero-point energy.

The electrum validum appears to be a self-organized structure akin to ball lightning (Singer, 1971, et al.) or the toroidal plasmoids produced by Bostick (1957). Could the anomalous persistence of ball lightning be associated with a zero point energy coherence as well? (Egely, 1986, Jennsion, 1990)

How plasmas could cohere the zero-point energy can be understood from examining the vacuum polarization description of the constituent particles (Rausher, 1968). Quantum electrodynamics shows that the vacuum polarization of electrons is quite different than atomic nuclei (Scheck, 1983, Reinhardt, et al., 1980). Electrons (especially when bound in matter) exhibit a random, cloud-like interaction with the vacuum fluctuations and are effectively in thermodynamic equilibrium with it. On the other hand, nuclei exhibit stable, orderly lines of vacuum polarization which converge radially onto the particle. This allows them to launch and respond to local vacuum polarization displacement currents, which conduction band electrons may not be able to detect. The abrupt motion of nuclei can locally drive the ZPE far from equilibrium and, when combined with the nonlinear dynamics of both the plasma and the ZPE, can fulfill Prigogine's conditions for system self-organization to trigger the formation of exotic, energetic vacuum states similar to those that sometimes occur in heavy-ion collision experiments (Celenza, 1986, et al.). Evidence for a ZPE coherence in plasmas arises from the observed anomalies associated with the ion-acoustic mode where the plasma's ions are undergoing synchronous oscillations. Here is observed runaway electrons (Kiwamoto, et al., 1979), anomalous heating (Sethian, 1978, et al.), and high frequency voltage spikes (Kalinin, 1970, et al.). In general, plasma physicists have not been looking for energy anomalies in their experiments, but recently Chernetskii (Samokhin, 1990) has claimed to produce more heat output in his plasma experiments than energy input and attributes the ZPE as the source of the excess energy.

Another class of experiments yielding anomalous heat are the "cold fusion" experiments of which there have recently been numerous successes (Storms, 1991). These experiments can manifest ion-acoustic type oscillations of the deuterons within the hydride lattice sites especially under the conditions of deuterium supersaturation where all the sites are occupied (and diffusion is inhibited). Here within a single crystalline grain of the hydride, the deuterons may undergo coherent oscillations creating a macroscopic ZPE vacuum polarization. If the ZPE is coherently interacting with the system, then it can trigger some fusion events (Jandel, 1990) as well as produce abundant heat without fusion. Especially relevant are the experiments of Mills and Kneizys (1991) who can generate anomalous heat with 100% repeatability in their light water experiments, and there has been reported numerous independent replications (Mallove, 1992, et al.). Also Dufour (1993) has created a simple experiment involving just sparking in hydrogen with a stainless steel electrode which produces excess heat with 100 percent repeatability. Pappas (1991) likewise claims excess energy from sparking. Recently Mizuno, et al. (1993) have investigated proton conductors and claim their experiment outputs 70,000 times more heat than accounted for by the input power. The zero-point energy just might be the source of the anomalous energy in experiments where nuclei motion is involved.

Another experimental area where energy anomalies could be associated with coherent ion motion occurs in sonoluminescence (Walton and Reynolds, 1984). Here ultrasonic excitation of water causes emission of blue light that is visible to the naked eye. It has previously been assumed that the collapse of cavitation bubbles produces a sufficiently abrupt heating that it can dissociate some water molecules into a plasma followed by chemiluminescence of its constituents. However, the recent research of Barber and Putterman (1991) have shown that the photon emissions are much too rapid for the fastest atomic electron transitions. Moreover, they show that the phenomenon represents a 10^{11} amplification of energy. Nobel laureate Schwinger (1993) has proposed a zero-point energy interaction to explain the source of energy for sonoluminescence.

Along similar lines, coherent nuclei coupling to the ZPE could be suggested to explain the anomalies associated with fractoemission (Preparata, 1991). The abrupt fracturing of crystals has yielded anomalously powerful light emissions as well as excessively accelerated electrons. Furthermore, the anomalous events could persist for hours. Preparata (1991) has proposed a theory of superradiance where coherent radiation from a plasma produced around the fracture is synchronous with the plasma oscillations and creates an excited, standing-wave, bound state within the fracture cavity. Preparata's model has the ZPE coherently interacting with plasma ion-acoustic oscillations that can manifest an energy concentration which produces the anomalously powerful events. The proposed ZPE coherence exhibits stability similar to ball lightning or Shoulder's (1991) electrum validum, and appears to explain the events' anomalous persistence.

Inventions

Evidence for tapping the zero-point energy as an energy source also comes from inventions whose principle of operation involves the same ion-acoustic type activity that exhibits energy anomalies in the experiments. For example in the 1930's Moray (1978) produced a solid state device that was well witnessed to output 50 kilowatts of electricity. Moray stressed that the basis of success was maintaining ion oscillations in the plasmas within the device's tubes. Along similar lines Brown (1989) has reproduced a resonant nuclear battery that is similar to the source (Krypton 85) to maintain a plasma which couples to a tuned circuit and outputs 5 watts. The one curie radioactive source could only output power at best (assuming 100% efficiency in converting its mass to energy) on the order of 5 milliwatts. However, if the tuned circuit resonates at the plasma's ion-acoustic frequency, and this mode couples coherently to the ZPE, it could explain the source of anomalous power.

The ZPE ion-acoustic hypothesis may be applied to explain anomalous energy production in the water dissociation inventions of Puharich (1981) and Meyer (1991). Both inventions excite water in specially shaped vessels with large voltage pulses designed to resonate the water molecules at the frequencies of the hydrogen-oxygen electron bonds. Puharich stresses the importance of triggering an acoustical resonance within the water molecule itself, and both inventors claim to require far less input energy to accomplish the dissociation than what is returned when the hydrogen is burned as a fuel. This work is reminiscent of Keely's research (Moore, 1971) who claimed to easily dissociate water when excited at the appropriately resonant ultrasonic frequencies. Davidson (1990) reports of an anomalously violent event when a column of water, adjusted to support acoustical standing waves, was driven with a barium titanate ultrasonic transducer at the water dissociation frequency identified by Keely (approximately 43 KHz). If the synchronous motions of the nuclei comprising the water molecules induce a coherenence in the ZPE, the anomalous energy produced in the resonant dissociation of water could be explained.

Another invention where resonant ionic motion may be launching anomalous energy is the battery pulsing circuit originally discovered by Bedini (1991) and investigated by Panici (1992). If a new lead acid battery is pulsed charged by a fast rise time square wave (unipolar positive above 20 volts, 50% duty cycle, and on the order of 300 to 400 Hz) anomalous power can sometimes be generated that seems to be related to a critical timing in the arrival of the pulse edges. The most surprising claim regarding this experiment is the production of "cold current" where appreciable output power (200 watts) can be guided along thin wires (no. 28) without heating them. The effect works only on new lead acid batteries; a battery that has been discharged and recharged by a standard DC battery charger will not produce it. The anomalous behavior could be explained as follows: The battery plates consist of porous lead or lead peroxide which exhibit a delicate dendritic structure much like the description of the palladium surface in the theories of electrochemistry (Storms, 1991, Gluck, 1993). During both the charge and discharge reactions, protons (hydrogen ions) form on the dendritic surface within the porous plates. The intense electric fields from the sharp pointed dendrites can accelerate the protons, and when stimulated by an appropriate pulsed excitation, the protons can undergo coherent, synchronous oscillations similar to those described for the cold fusion experiments. The oscillations may be resonant at a higher frequency than the square wave excitation, but by adjusting the square wave to an exact subharmonic, the pulse edges can arrive in phase with the resonance. Discharging the battery accumulates sulfate on the porous plates, and recharging it with DC ruins the original dendritic surface (Barak, 1980). Thus a new battery must be used for this experiment.

The surprising cold current effect has been associated with other inventions (e.g. Moray (1978) demonstrated it and used no. 30 wire within his 50 KW device). It might be explained by the hypothesis that the coupling of synchronous ion oscillations with the ZPE can manifest macroscopic vacuum polarization displacement currents (King, 1984) which can surround a conductive wire and be guided by it. Conduction band electrons, lacking stable, radially convergent vacuum polarization lines, have a minimal interaction with these displacement currents, and the wires remain cool. If engineers could replicate this effect, zero-point energy research would receive widespread interest.

Another area where coherent ion motion is associated with energy anomalies involves abrupt electrical discharges. Graneau (1985) has experimentally demonstrated that an anomalous force is associated with a large, abrupt electrical discharge in water. He observed a threshold effect where the same amount of energy is discharged from a capacitor bank into a water vessel, but with different rise times. At slow rise times, no motion is observed, but at a threshold where the pulse becomes abrupt enough, the water exhibits an explosive expansion. Johnson (1992) surmises from his experiments that the force is related to current density. Along similar lines the engines of Gray (1976) and Papp (1984) utilize abrupt electrical discharges to produce an anomalous driving force as well as excessive output power. The hypothesis of coherent ion motion coupling to the ZPE seems to likewise fit these inventions.

Pappas (1991) has proposed from his derivation from Ampere's law that a powerful enough electric discharge, where the electron velocities exceed 70% the speed of light, can produce anomalous energy when the discharge strikes the anode plate. An alternative, similar hypothesis is that the powerful discharge hitting the anode can produce a plasma ionic surge that, if coupled coherently with the ZPE, would manifest a greater voltage spike than would be otherwise predicted by classical theory. Many other inventors have also claimed that the anomalous energy production in their devices takes the form of abrupt voltage spikes, and they would have a practical device if only they could efficiently convert this form of energy.

Hyde (1990) was faced with a similar problem of absorbing large voltage pulses produced in his invention. He solved the problem and created an advanced embodiment which produced over 20 KW of net output power while free running. Hyde designed a solid state voltage divider that takes advantage of the abrupt rise time of the incident pulses, and it is disclosed in his patent.

The voltage divider circuit diagramed in the patent (Figure 1 here) has been questioned by electrical engineers since the line (herein labeled L_1) connecting the positive sides of the capacitors (labeled 100, 86 and 106) as well as the line (herein labeled L_2) connecting the negative sides of these capacitors (thus connecting them in parallel) essentially short out the series connections provided by the diodes 108 and 104. If the diagram is analyzed as a lumped circuit, the diodes 108 and 104 would always be back biased and appear like open switches, and the circuit could be viewed (when the capacitors are combined) as equivalent to a standard rectifier. Thus when analyzed as a lumped circuit, the diagramed circuit appears unable to provide voltage division.

However, analyzing the diagram as a distributed circuit or transmission line suggests that it could produce voltage division of sharp pulses propagating through the circuit. To accomplish voltage division (with concomitant current multiplication), the capacitors must be charged in series and then discharged in parallel. If the electrical lines L_1 and L_2 are made considerably longer than the series path through the capacitors made considerably longer than the series path through the capacitors (100, 86, 106), then their residual line inductance would result in a higher (surge) impedance to sharp voltage pulses, and tend to channel the pulses through the (lower impedance) series path. If necessary, the inductance of L_1 and L_2 can be increased by shaping the wires into loose coils or by adding ferrite loading around the wires. Hyde developed the circuit empirically as he was attempting to control and absorb the very large and sharp voltage pulses generated in his device. The sharp pulses take the series path to charge the capacitors, and then the energy is discharged smoothly through the parallel path. Only by analyzing the diagram as a distributed circuit can its voltage division action be understood.

Note the diagram shows only a three stage illustration of a voltage division transmission line for sharp pulses. Clearly more stages can be used if further voltage division (and current multiplication) is needed. Hyde's circuit may be of value to inventors whose devices produce high voltage spikes from ion-acoustic activity. It might facilitate a very straight

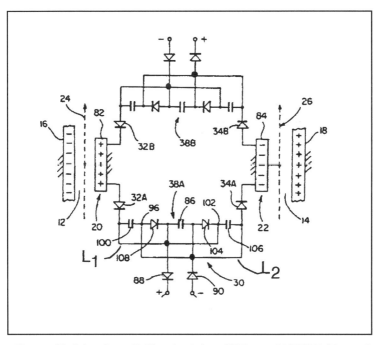

Figure 1 Hyde's voltage divider circuit from US Patent #4,897,592 (Figure 6)

forward, solid state method for tapping the zero-point energy.

A solid state method for cohering the ZPE might be achieved by inducing appropriate nuclei motion within materials that exhibit strong electrical to acoustical coupling. For example, the following means of electrical excitation can stimulate acoustical spinor waves in a highly polarizable dielectric such as barium titanate.

Excitor electrodes are mounted as shown (Figure 2) on a cylinder of barium titanate 45 degrees from the axis of maximum polarization. The electrode pairs A_1, A_2 and B_1, B_2 are alternately excited by an oscillating, unipolar waveform which causes the polarization vector to flex between the A and B directions. The oscillating polarization impresses an oscillating torque on the dielectric lattice which twists it to produce synchronous cycloidal movement of the lattice nuclei. At the appropriate distance along the cylinder above the first electrode group (or polarization layer) is another set of electrodes whose voltage is timed to drive the polarization vector in opposite phase; i.e., when the first layer is polarized in the A direction, the next layer should be polarized in the B direction and visa versa. In like fashion, more layers of alternating excitation electrodes can be mounted along the cylinder. The optimal distance between the layers will be related to the wavelength of the acoustical excitation whose drive frequency should be chosen to maximize the amplitude of the acoustical activity. The back and forth twisting action of the lattice can excite and maintain two counter-rotating, acoustical spinor waves. If there is a direct coupling of the synchronous nuclei motion to the ZPE, there could then arise dual, counter-rotating, displacement current vortices driven by a spatially coherent, dynamic vacuum polarization. Such dual vortices would be a macroscopic formation analogous to virtual particle pair production. Quantum electrodynamics requires that any form arising from the ZPE be created in opposing pairs in order to conserve charge, momentum and spin (angular momentum). If, instead of a linear cylinder, the dielectric is shaped as a toroid, the dual spinor waves can close into standing waves, and a significant ZPE coherence might thus be induced within a solid state structure.

In a similar fashion, acoustical spinor waves can be created within a highly permeable magnetic material such as barium ferrite by using an equivalent magnetic field excitation in place of the electrodes. The oscillating magnetic vectors of the shifting domains couple to the ferrite lattice and elastically twist it back and forth (Cieplak, 1980). Experimental evidence that such activity could cohere the zero-point energy appears to be exhibited by the energy producing invention of Sweet (1991).

Vortices and Precession

The greatest zero-point energy coherence may arise in experiments creating a vortex or precessional flow of charged fluid or plasma. Schauberger (Alexandersson,1990, et al.) did a series of experiments circulating water in specially spiraled pipes. At certain velocities the fluid flow manifested negative resistance, i.e. energy creation, as well as a strange, bluish glow appearing in the water near the bottom of the vortex. Such a glow is reminiscent of sonoluminescence (discussed previously). Also Chernetskii (Samokhin, 1990) observed that the greatest output of anomalous energy in his plasma experiments was associated with the cycloid motion of the plasma particles. Such spiraling plasma particle motion is also the key operating point for the energy invention by Spence (1988). The theoretical research by Reed (1992), Jennison (1978) and Winter (1991) suggests a golden mean, logarithmic spiral is the three-space projection of a fundamental, hyperspatial flow of an ether (or ZPE flux). Systems whose dynamics align with this flow tap the zero-point energy.

An aligned interaction with the zero-point energy flux may alter local gravity. Laithwaite (Davidson, 1989) observed in his experiments that a gyroscope displaced along a particular spiraling path while precessing would manifest a gravitational and inertial anomaly. Puthoff (1989) suggests that gravity is derived from the action of the ZPE. This view is essentially supported by general relativity since the ZPE must be part of the stress-energy tensor which curves the space-time metric. If a system induces coherence in the ZPE, a gravitational anomaly or a change in the local pace of time could manifest (Kozyrev, 1968).

Precessional motion may be the method to manifest the largest ZPE coherence. DePalma (1973) observed a direct gravitational and inertial anomaly in his experiments involving forced precession of a counter-

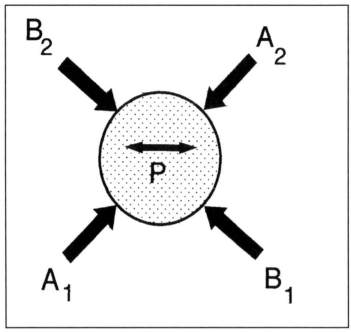

Figure 2 Cross section of barium titanate cylinder showing placement of excitor electrodes A_1, A_2, B_1, B_2 and axis of maximum polarization P.

rotating pair of gyroscopes. Schauberger used twisted pipes of an egg shaped oval cross section to induce an inner spin so that the spiral pipe formation would cause precessional motion in the water flow. A natural form to induce precessional flow is a helix tube closed to form a vortex ring. This shape was observed in the plasmoids by Bostick (1957) as well as that suggested for ball lightning (Johnson, 1965). Roden's (1992) theoretical treatise features a precisely shaped spiral circulating poloidally around a toroid. Medvedeff (1961) has the vortex ring as the basis for the elementary particles, Honig (1986) uses it to model the photon, and Winterberg (1990) has it as the basis for a unified field theory. Childress (1991) describes the vimanas, ancient flying ships reported in the Hindu vedic texts which had a propulsion system based on circulating mercury through a vortex ring. A surprisingly large gravitational anomaly might well be created by rapidly pumping positively charged fluid or plasma through a vortex ring piping system. Such an experiment might well be robust, and repeat reliably for all investigators.

Scalar Fields

Another method for cohering the zero-point energy involves abruptly bucking, electromagnetic (EM) fields. When EM fields are in perfect opposition, the field vectors exactly cancel creating a zero resultant. However, there still exists a stress in the fabric of space and it manifests as a scalar EM potential. Aharonov and Bohm (1959, et al.) have shown that the EM potential affects the phase of the quantum mechanical wave function associated with the elementary particles. Bearden (1986) has emphasized that the resultant stress is actually a coherence in the ZPE and can propagate as scalar waves. King (1986) has suggested a hyperspatial model for such a coherence involving vortex rings. The model is based on geometrodynamics (Wheeler, 1962) where the ZPE manifests as a electric flux flowing orthogonally to our three space. Abruptly opposing EM field transients pinch and release this flux which tend to guide (or orthorotate) some of the flow into our three space. Timing is critical: The quicker and more abrupt the opposing EM field transients, the greater the flux orthorotation.

An experiment to demonstrate the effect of abruptly bucking electromagnetic fields involves the use of a caduceus coil (Smith, 1964). The caduceus coil has two windings of opposite helicity criss-crossing like the snakes on the caduceus staff. Perfect symmetry is important, for the coils are driven simultaneously by abrupt electric pulses. As the they propagate up both windings, the pulse edges must remain perfectly aligned to maintain the bucking field action. The sharper the rise time of the aligned pulse edges, the greater the effect. The coil should be wound on a hollow tube in order to facilitate experiments with a variety of core materials. Burridge (1979) describes the use of ferrite cores, and Van Tassel (Dollard, 1988) experimented with quartz crystal cores. A synergistic ZPE coherence might well be induced by having the core be comprised of any of the previously suggested apparatus for cohering the ZPE. For example, a long plasma tube excited at its ion-acoustic resonance, or a tube in which there occurs a plasma vortex or other plasmoids such as produced by Shoulders (1991). Meyer (1989) achieved considerable success by combining many ideas, and a pulsed caduceus coil excitation may well augment any device that is cohering the zero-point energy.

Summary

From the concentrated vacuum polarization of atomic nuclei arises the possibility of triggering a macroscopic ZPE coherence with the synchronous motion of many nuclei. Experimental evidence arises from energy anomalies associated with heavy-ion collisions, plasma ion-acoustic oscillations, sonoluminescence, fractoemission, large charge density plasmoids, ball lightning, abrupt electric discharges, and light water "cold fusion" experiments. Further supporting evidence arises from inventions that utilize coherent ion motions and output excessive energy. The sharp voltage spikes that arise from ion-acoustic activity can be converted by use of circuits such as Hyde's voltage divider. Large effects might be produced by solid state methods that manifest dual, counter-rotating, acoustical, lattice spinor waves. Precessing charged plasma or fluid by pumping it through a helical vortex ring might produce a sufficient ZPE coherence to manifest a gravitational anomaly. Abruptly bucking electromagnetic fields from a pulsed caduceus coil could further stimulate a coherent ZPE interaction with whatever is placed within its core. If synchronous ion-acoustic activity coheres the zero-point energy, there will be many more inventions forthcoming, and a new energy source will be recognized.

References

Aharonov, Y. and D. Bohm (1959), "Significance of Electromagnetic Potentials in the Quantum Theory," Phys. Rev. 115(3), page 485; Olariv, S. and I.I. Popescu (1985), "The Quantum Effects of Electromagnetic Fluxes," Rev. Mod. Phys. 57(2), pages 339-436.

Alexandersson, O.(1990), Living Water: Viktor Schauberger and the Secrets of Natural Energy, Gateway Books, Bath, UK. Also Frokjaer-Jensen, B.(1981), "The Scandinavian Research Organization and the Implosion Theory (Viktor Schauberger)," Proc. First International Symposium on Nonconventional Energy Technology, Toronto, pages 78-96.

Barak, M.(1980), Electrochemical Power Sources, I.E.E. and Peter Peregrinus LTD., NY, pages 188-190.

Barber, B.P. and S.J. Putterman (1991), "Observation of synchronous picosecond sonoluminescence," Nature 353, pages 318-320.

Bearden, T.E.(1986), Fer-De-Lance: A Briefing on Soviet Scalar Electromagnetic Weapons, Tesla Book Co., Millbrae, CA., pages 107-108.

Bedini, J.(1991), "The Bedini Free Energy Generator," Proc. 26th IECEC vol. 4, pages 451-456.

Bostick, W.H.(1957), "Experimental Study of Plasmoids," Phys. Rev. 106(3), page 404.

Boyer, T.H.(1975), "Random Electrodynamics: The theory of classical electrodynamics with classical electromagnetic zero-point radiation," Phys. Rev. D 11(4), pages 790-808.

Boyer, T.H. (1976), "Equilibrium of random classical electromagnetic radiation in the presence of a nonrelativistic nonlinear electric dipole oscillator," Phys. Rev. D 13(10), pages 2832-45.

Brown, P.M.(1989), "Apparatus for Direct Conversion of Radioactive Decay Energy to Electrical Energy," U.S. Patent No. 4,835,433; ... (1987), "The Moray Device and the Hubbard Coil were Nuclear Batteries," Magnets 2(3), pages 6-12; ... (1990), "The Resonant Nuclear Battery," International Tesla Symposium, Colorado Springs.

Burridge, G.(1979), "The Smith Coil," Psychic Observer 35(5), pages 410-416.

Bush, R.T. (1992), "A Light Water Excess Heat Reaction Suggests that 'Cold Fusion' is 'Alkali-Hydrogen Fusion'," Fusion Tech.22, page 287.

Fundamentals of a Zero-Point Energy Technology

Celenza, L.S. and V.K. Mishra, C.M. Shakin, K.F. Liu (1986), "Exotic States in QED," Phys. Rev. Lett. 57(1), page 55; Caldi, D.G. and A. Chodos (1987), "Narrow e⁺e⁻ peaks in heavy-ion collisions and a possible new phase of QED," Phys. Rev. D 36(9), page 2876; Jack Ng, Y. and Y. Kikuchi (1987), "Narrow e⁺e⁻ peaks in heavy-ion collisions as possible evidence of a confining phase of QED," Phys. Rev. D 36(9), page 2880; Celenza,L.S. and C.R. Ji, C.M. Shakin (1987), "Nontopological solitons in strongly coupled QED," Phys. Rev. D 36(7), pages 2144-48.

Childress, D.H.(1991), Vimana Aircraft of Ancient India and Atlantis, Adventures Unlimited Press, Stelle, IL.

Cieplak, M. and L.A. Turski (1980), "Magnetic solitons and elastic kink-like excitations in compressible Heisenberg chain," J. Phys. C: Solid State Physics 13, pages L 777-780.
Cole, D.C. and H.E. Puthoff (1993), "Extracting energy and heat from the vacuum," Phys. Rev. E 48(2), pages 1562-65.

Davidson, D.A.(1990), Energy: Breakthroughs to New Free Energy Devices, Rivas, Greenville, TX, pages 16-18.

Davidson, J.(1989), The Secret of the Creative Vacuum, C.W. Daniel Co. Ltd., Essex, UK, pages 258-262.

DePalma, B.E. and C.E. Edwards (1973), "The Force Machine Experiments," privately published.

Dirac, P.A.(1930), Roy. Soc. Proc.126, page 360. Also Gamow, G.(1966), Thirty Years that Shook Physics, Double Day, NY.

Dollard, E.(1988), "Van Tassel's Caduceus Coils," private communication. Van Tassel experimented with numerous caduceus coils that often contained quartz cores. His notes stated that the cross-over angle for the two opposing windings should be 22.5 degrees.

Dufour, J.(1993), "Cold Fusion by Sparking in Hydrogen Isotopes," Fusion Technology 24, pages 205-228.

Egely, G.(1986), "Energy Transfer Problems of Ball Lightning," Central Research Institute for Physics, Budapest, Hungary.

Forward, R.L.(1984), "Extracting electrical energy from the vacuum by cohesion of charged foliated conductors," Phys. Rev. B 30(4), pages 1700-2.

Gluck, P.(1992), Letters to the editor, Fusion Facts 4(7), pages 22-24.

Graneau, P. and P.N. Graneau (1985), "Electrodynamic Explosions in Liquids,"

Appl. Phys. Lett. 46(5), pages 468-470.

Gray, E.V.(1976), "Pulsed Capacitor Discharge Electric Engine," U.S. Patent No. 3,890,548.

Haisch, B. and A. Rueda, H.E. Puthoff (1994), "Inertia as a zero-point field Lorentz force," Phys. Rev. A 49(2), pages 678-694.

Hathaway, G.(1991), "Zero-Point Energy: A New Prime Mover? Engineering Requirements for Energy Production & Propulsion from Vacuum Fluctuations," Proc. 26th IECEC vol. 4, pages 376-381.

Hiller, R. and S. Putterman, B. Barber (1992), Phys. Rev. Lett. 69, pages 1182-84.

Honig, W.M.(1986), The Quantum and Beyond, Philosophical Library, NY.

Hyde, W.W.(1990), "Electrostatic Energy Field Power Generating System," U.S. Patent No. 4,897,592. The invention is summarized in King (1991).

Jandel, M.(1990), "Cold Fusion in a Confining Phase of Quantum Electrodynamics," Fusion Tech. 17, pages 493-499.

Jennison, R.C.(1978), "Relativistic Phase-Locked Cavities as Particle Models," J. Phys. A Math Gen. VII(8).

Jennison,R.C.(1990), "Relativistic Phase-Locked Cavity Model of Ball Lightning," Electronics Laboratory, University of Kent, U.K.

Johnson, G.L.(1992), "Electrically Induced Explosions in Water," Proc. 27th IECEC vol. 4, pages 4.335-338.

Johnson, P.O.(1965), "Ball Lightning and Self Containing Electromagnetic Fields," Am. J. Phys. 33, page 119.

Kalinin, Yu G. et al.(1970), "Observation of Plasma Noise During Turbulent Heating," Sov. Phys. Dokl. 14(11), page 1074; Iguchi, H.(1978), "Initial State of Turbulent Heating of Plasmas," J. Phys. Soc. Jpn. 45(4), page 1364; Hirose, A.(1974), "Fluctuation Measurements in a Toroidal Turbulent Heating Device," Phys. Can. 29(24), page 14.

King, M.B.(1984), "Macroscopic Vacuum Polarization," Proc. Tesla Centennial Symposium, International Tesla Society, Colorado Springs, pages 99-107. Also (1989), pages 57-75.

King, M.B.(1986), "Cohering the Zero-Point Energy," Proc. of the 1986 International Tesla Symposium, Colorado Springs,section 4, pages 13-32. Also (1989), pages 77-106.

King, M.B.(1989), Tapping the Zero-Point Energy, Paraclete Publishing, Provo, UT.

King, M.B.(1991), "Tapping the Zero-Point Energy as an Energy Source," Proc. 26th IECEC vol.4, pages 364-369.

Kiwamoto, Y. and H. Kuwahara, H. Tanaca (1979), "Anomalous Resistivity of a Turbulent Plasma in a Strong Electric Field," J. Plasma Phys. 21(3), page 475.

Kozyrev, N.A.(1968), "Possibility of Experimental Study of the Properties of Time," Joint Publication Research Service, Arlington VA.

La Violette, P.A.(1985), "An introduction to subquantum kinetics...," Intl. J. Gen. Sys. 11, pages 281-345; ... (1991), "Subquantum Kinetics: Exploring the Crack in the First Law," Proc. 26th IECEC vol. 4, pages 352-357.

Mallove, E.F.(1992), "Protocols for Conducting Light Water Excess Energy Experiments," Fusion Facts 3(8), page 15; Noninski, V.C.(1992), "Excess Heat during the Electrolysis of a Light Water Solution of K_2CO_3 with a Nickel Cathode," Fusion Tech. 21, pages 163-167.

Medvedeff, N.J.(1961), Nuclear Dynamics, privately published, Hanover, MA.

Meyer, S.L.(1991), The Birth of a New Technology, Water Fuel Cell, Grove City, OH; ... (1989), "Controlled Process for the Production of Thermal Energy from Gases and Apparatus Useful Therefore," U.S. Patent No. 4,826,581; ... (1990), "Method for the Production of a Fuel Gas, (Electrical Polarization Process)," U.S. Patent No. 4,936,961.

Mills, R.L. and S.P. Kneizys (1991), "Excess Heat Production by the Electrolysis of an Aqueous Potassium Carbonate Electrolyte and the Implications for Cold Fusion," Fusion Tech. 20, pages 65-81.

Milonni, P.W. and R.J. Cook, M.E. Goggin (1988), "Radiation pressure from the vacuum: Physical interpretation of the Casimir force," Phys. Rev. A 38(3), pages 1621-23.

Mizuno, T. and M. Enyo, T. Akimoto, K. Azumi (1993), "Anomalous Heat Evolution from $SrCeO_3$ - Type Proton Conductors during Absorption/Desorption of Deuterium in Alternate Electric Field," 4th Int. Conf. on Cold Fusion. Abstract in Fusion Facts, Dec. 1993, page 30.

Moore, C.B.(1971), Keely and His Discoveries, Health Research, Mokelumne Hill, CA.

Moray, T.H. and J.E. Moray (1978), The Sea of Energy, Cosray Research Institute, Salt Lake City.

Panici, D.(1992), private communication.

Papp, J.(1984), "Inert Gas Fuel, Fuel Preparation Apparatus and System for Extracting Useful Work from the Fuel," U.S. Patent No. 4,428,193.

Pappas, P.T.(1991), "Energy Creation in Electrical Sparks and Discharges: Theory and Direct Experimental Evidence," Proc. 26th IECEC vol. 4, pages 416-423.

Preparata, G.(1991), "A New Look at Solid-State Fractures, Particle Emission and Cold Nuclear Fusion," Il Nuovo Cimento 104 A(8), page 1259; ... (1990), "Quantum field theory of superradiance," in Cherubini, R., P. Dal Piaz, B. Minetti (editors), Problems of Fundamental Modern Physics, World Scientific, Singapore.

Prigogine, I. and G. Nicolis (1977), Self Organization in Nonequilibrium Systems, Wiley, NY; Prigogine, I. and I. Stengers (1984), Order Out of Chaos, Bantam Books, NY.

Puharich, A.(1981), "Water Decomposition by Means of Alternating Current Electrolysis," Proc. First International Symposium on Nonconventional Energy Technology, Toronto, pages 49-77; ... (1983), "Method and Aparatus for Splitting Water Molecules," U.S. Patent No. 4,394,230.

Puthoff, H.E.(1987), "Ground state of hydrogen as a zero-point fluctuation determined state," Phys. Rev. D 35(10), pages 3266-69.

Puthoff, H.E.(1989), "Gravity as a Zero-Point Fluctuation Force," Phys. Rev. A 39(5), page 2333.

Puthoff, H.E.(1990), "The energetic vacuum: implications for energy research," Spec. Sci. Tech. 13(4), pages 247-257.

Reed, D. (1992), "Toward a Structural Model for the Fundamental Electrodynamic Fields of Nature," Extraordinary Science IV(2), pages 22-33; ... (1993), "Evidence for the Screw Electromagnetic Field in Macro and Microscopic Reality," Proc. Int. Sym. on New Energy , pages 497-510; ... (1994), "Beltrami Topology as Archetypal Vortex," Proc. Int. Sym. on New Energy, pages 585-608.

Reinhardt, J. and B. Muller, W. Greiner (1980), "Quantum Electrodynamics of Strong Fields in Heavy Ion Collisions," Prog. Part. and Nucl. Phys. 4, page 503.

Rausher, E.A.(1968), "Electron Interactions and Quantum Plasma Physics," J.

Fundamentals of a Zero-Point Energy Technology

Plasma Phys. 2(4), page 517.

Roden, M.(1992), Aero Dynamics, Aero Dynamics Master Society, San Ysidro, CA.

Samokhin, A.(1990), "Vacuum energy - a breakthrough?" Spec. Sci. Tech. 13(4), page 273.

Scheck, F.(1983), Leptons, Hadrons and Nuclei, North Holland Physics Publ., NY, pages 213-223.

Schwinger, J.(1993), "Casimir light: The source," Proc. Natl. Acad. Sci. USA 90, pages 2105-6.

Scully, M.O. and M. Sargent (March 1972), "The Concept of the Photon," Physics Today, page 38.

Senitzky, I.R.(1973), "Radiation Reaction and Vacuum Field Effects in Heisenberg-Picture Quantum Electrodynamics," Phys. Rev. Lett. 31(15), page 955.

Sethian, J.D. and D.A. Hammer, C.B. Whaston (1978), "Anomalous Electron-Ion Energy Transfer in a Relativistic-Electron-Beam Heated Plasma," Phys. Rev. Lett. 40(7), page 451; Robertson, S. and A. Fisher, C.W. Roberson (1980), "Electron Beam Heating of a Mirror Confined Plasma," Phys. Fluids 32(2), page 318; Tanaka, M. and Y. Kawai (1979), "Electron Heating by Ion Acoustic Turbulence in Plasmas," J. Phys. Soc. Jpn. 47(1), page 294.
Shoulders, K.R.(1991), "Energy Conversion Using High Charge Density," U.S. Patent No. 5,018,180.

Singer, S.(1971), The Nature of Ball Lightning, Plenum Press, NY; Silberg, P.A.(1962), "Ball Lightning and Plasmoids," J. Geophys. Res. 67(12), page 4941.

Smith, W.B.(1964), The New Science, Fern-Graphic Publ., Mississauga, Ontario.

Spence G.M.(1988), "Energy Conversion System," U.S. Patent No. 4,772,816.

Storms, E.(1991), "Review of Experimental Observations about the Cold Fusion Effect," Fusion Tech. 20, pages 433-477.

Suzuki, M.(1984), "Fluctuation and Formation of Macroscopic Order in Nonequilibrium Systems," Prog. Theor. Phys. Suppl. 79, pages 125-140; Hasegawa, A.(1985), "Self-Organization Processes in Continuous Media," Adv. Phys. 34(1), pages 1-41; Firrao, S.(1984), "Physical Foundations of Self-Organizing Systems Theory," Cybernetica 17(2), pages 107-124; Haken, H.(1971), Synergetics, Springer Verlag, NY.

Sweet, F. and T.E. Bearden (1991), "Utilizing Scalar Electromagnetics to Tap

Vacuum Energy," Proc. 26th IECEC vol. 4, pages 370-375; ... (1988), "Nothing is Something: The Theory and Operation of a Phase-Conjugate Vacuum Triode," private communication.

Walton, A.J. and G.T. Reynolds (1984), "Sonoluminescence," Adv. Phys. 33(6), pages 595-600.

Wheeler, J.A.(1962), Geometrodynamics, Academic Press, NY.

Winter, D.(1991), The Star Mother, Geometric Keys to the Resonant Spirit of Biology, Crystal Hill Farm Publications, Eden, NY.

Winterberg, F.(1990), "Maxwell's Equations and Einstein-Gravity in the Planck Aether Model of a Unified Field Theory," Z. Naturforsch. 45 a, pages 1102-16; ... (1991), "Substratum Interpretation of the Quark-Lepton Symmetries in the Planck Aether Model of a Unified Field Theory," Z. Naturforsch. 46 a, pages 551-559.

Fundamentals of a Zero-Point Energy Technology

VACUUM ENERGY VORTICES

April 1994

Abstract

Abrupt motion or helical motion of atomic nuclei can cause a coherent, multibody vacuum polarization in the zero-point energy (ZPE). Supporting evidence arises from inventions and experiments that produce anomalous energy from events such as abrupt electrical discharges, plasma ion-acoustic oscillations, sonoluminescence and fractoemission. Excess energy is produced when ions precess through a force-free vortex. Counter-rotating plasmas might create pair production of macroscopic ZPE displacement current vortices. Examples include the Swiss ML Converter and Sweet's "Vacuum Triode Amplifier" where twisting lattice grains within a barium ferrite ceramic apparently induce fractoemission.

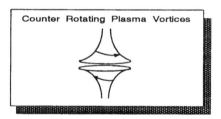

Counter Rotating Plasma Vortices

Introduction

Modern physics views the vacuum of empty space not as a void but as a plenum of randomly fluctuating electromagnetic fields known as the zero-point energy (ZPE). These vacuum fluctuations persists at zero degrees Kelvin, and many physicists have shown they possess an enormous energy density (Hathaway, 1991). For example Wheeler (1962) derives a density of 10^{94} grams/cm^3 for individual fluctuations on the scale of a Planck length (10^{-33} cm). The energy is not readily noticed at the classical level since the fluctuations are ubiquitous, and a standard radiation detector requires an energy difference to make a measurement. Nonetheless there is a point of view well represented in the physics literature that shows the ZPE is the basis of the material world. Boyer (1975) shows how the blackbody radiation spectrum as well as other quantum events is attributed to the ZPE. Senitsky (1973) has suggested that an elementary particle's very existence is intertwined with the ZPE. Puthoff (1987) has shown the hydrogen atom's stability is due to a ZPE interaction which prevents the electron from collapsing into the nucleus. Puthoff (1989) has also shown how gravity can be derived from the action of the ZPE, and recently Haisch, Rueda and Puthoff (1994) have shown that the ZPE could be the basis of inertia as well. There have also been proposed ZPE models for the photon (Scully, 1972, Honig, 1986) and the electron (Jennison, 1978). The zero-point energy might well be the primary underpinning of all physics.

Can the zero-point energy be tapped as an energy source? At first this idea seems to be a blatant violation of conservation of energy. But if the ZPE is physically real, then there is energy available and its conservation would not be the issue. The real issue centers on the second law of thermodynamics, the law of entropy, for how could a system based on chaotic energy fluctuations evolve into coherence? Prigogine (1977) won the Nobel prize in chemistry for showing how a system can evolve from chaos into order. The system must exhibit three characteristics: 1) It must be nonlinear, 2) far from equilibrium and 3) have an energy flux through it. The published theories of the ZPE fulfill these conditions (King, 1989, 1993). Recently Cole and Puthoff (1993) published a proof showing that, in principle, the ZPE can be tapped as an energy source without violating thermodynamics. Forward (1984) has devised a simple battery based on the Casimir force (Milonni, 1988) where a ZPE radiation pressure squeezes charged plates together. Also, Boyer (1976) has mathematically shown that a system consisting of a nonlinear dipole can absorb energy from particular modes of the ZPE spectrum. As chaos theory and system theory are applied to the ZPE, new models of matter, fields and spacetime arise (e.g. LaViolette, 1985, Winterberg, 1990). Combining the theories of system self-organization (e.g. Suzaki, 1984, et al.) with the theories of the zero-point energy open the possibility for a new energy source.

Ion Motion

How can an engineer design a system that will tap the zero-point energy? Quantum electrodynamics (QED) shows how the various elementary particles interact with the ZPE via vacuum polarization (Scheck, 1983, Reinhardt, 1980). Electrons, especially when in conductors, exhibit a random vacuum polarization where the electron cloud is essentially in thermodynamic equilibrium with the ZPE fluctuations. Thus normal antennas, conductors and electron circuits would not experience any net ZPE interaction. On the other hand the atomic nuclei exhibit an orderly vacuum polarization with polarization lines converging steeply onto the particle. The abrupt motion of nuclei or ions would be expected to induce a cohering effect on the ZPE, and this has been observed in accelerator experiments involving heavy ion collisions (Celenza, 1986, et al.). The ZPE coherence would be greater if a large number of nuclei (or ions) undergo synchronous or abrupt motion. This behavior occurs in plasmas during

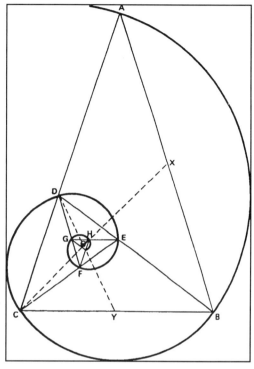

Figure 1. Logarithmic spiral with embedded phi proportioned triangles.

ion-acoustic resonance, and it has manifested energetic anomalies such as high frequency voltage spikes (Lalinin, 1970, et al.), run away electrons (Kiwamoto, 1979) and anomalous heating (Sethian, 1978, et al.). These anomalies might be evidence for a macroscopic ZPE coherence.

There are inventions and experiments producing excessive energy that have nuclei or ion motion at their basis. In the 1930's T. H. Moray (1978) produced a 50 kilowatt, free running, energy machine that relied on ion oscillations in its plasma tubes. Brown (1989) may have replicated the essence of Moray's discovery with his resonant nuclear battery. Brown uses a weak radioactive source to maintain a plasma which couples to a resonant circuit tuned at the plasma's ion-acoustic frequency. Also, the "cold fusion" experiments (Storms, 1991) invoke coherent nuclei motion. In the hydride lattice the deuterons (or protons) settle into shallow potential wells where they are free to oscillate or diffuse.

When the lattice is saturated, all the sites are occupied inhibiting the diffusion and allowing the nuclei to undergo synchronous oscillations. A ZPE coherence might be the primary source of anomalous energy in the cold fusion experiments (Jandel, 1990) since not only are there insufficient fusion by-products (neutrons, tritium, helium, etc.) but excess heat is produced repeatably in a class of light water experiments (Mills, 1991, Mallove, 1992, et al.). Moreover, Dufour (1993) has created a simple experiment involving just sparking in hydrogen with a stainless steel electrode, and it produces excess heat with 100 percent repeatability. Pappas (1991) likewise claims excess energy from sparking. Recently Mizuno, et al. (1993) have investigated proton conductors and claim their experiment outputs 70,000 times more heat than accounted for by the input power. The zero-point energy just might be the source of the anomalous energy in experiments where nuclei motion is involved.

Plasma Vortices

Vortical motion of plasma nuclei appear to produce even bigger effects. Chernetski (Samokhin, 1990, Michrowski, 1993) has observed anomalously excessive energy production in his plasma experiments and associates the greatest output with cycloidal motion of the plasma particles. Similarly Spense (1988) has based his energy invention on the plasma spiral. Reed (1992, 1994) has overviewed vortex theories in the physics literature and has concluded a particular form known as the "force free vortex" is a likely candidate for maximizing a ZPE interaction. The force free vortex originally arose in hydrodynamics, and is described by the equation:

$$\nabla \times v = kv$$

where v = the fluid velocity vector and k = a constant.

The curl of the fluid flow is in the same direction as the flow itself. The name "force free" applies because a fluid can spiral in this particular manner without being constrained by an external force. The curl equation gives rise to a set of helical solutions where the pitch of the helix is fixed and the radius may vary. Solutions are allowed where the radius gradually shrinks and the velocity increases concentrating energy toward the vortex tip. Reed also notes that the force free vortex occurs in plasma electrodynamics and is described by the curl of the magnetic field, **B**:

$$\nabla \times B = kB$$

Bostick (1966) has observed the formation of the plasma vortices in his experiments and notes that they tend to arise in symmetric pairs exhibiting opposite helicity. The plasma vortex helix can also close onto itself creating a vortex ring plasmoid. Bostick (1957) has also created these in his experiments and describes a "quantum condition" or integer ratio of the toroidal radius to the poloidal radius. Plasmoids also tend to be produced in pairs of opposite helicity and might well be an archetype illustration of electron-positron pair production arising from the QED vacuum. Ball lightning (Singer, 1971) has been modeled as a vortex ring plasmoid (Johnson,1965, Jennison, 1990), and its anomalous persistence might be indicative of a coherent ZPE interaction (Egely, 1986, King, 1989). A vortex geometry for plasma ion motion might contribute significantly in cohering the zero-point energy.

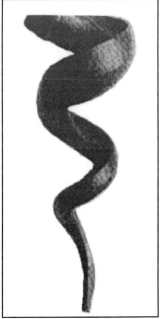

The force free vortex may be one constraint to the appropriate vortex geometry for an optimal ZPE coherence, but since the helix radius may vary arbitrarily, there are too many solutions. Another boundary condition is required to give a unique, single shaped vortex. The logarithmic spiral (Figure 1) might be a good candidate since it often appears in nature, and is associated with self-organization and growth (Huntley, 1970). The spiral is based on the ratio of the golden mean, phi (=1.618...), and has a fractal characteristic in that geometric forms imbedded within the spiral maintain their shapes. Pumping charged plasma or fluid through a pipe shaped as a force free vortex whose ra-

Figure 2. Schauberger's implosion vortex pipe.

dius contracts along the logarithmic spiral might align with an archetypal geometry toward which the ZPE would naturally self-organize.

Perhaps the most unusual energy anomalies associated with vortex experiments were claimed by Schauberger (Alexandersson, 1990, et al.). He empirically designed a particular tapered vortex based on his observations of turbulent water flow in nature. Schauberger pumped water through a pipe that was twisted into a precisely tapered helix whose cross section was also shaped to induce the water to naturally undergo a second order precessional spin as it flowed through the pipe toward the vortex tip. Normally it is difficult to force matter to undergo precession, yet when it does, there have been anomalies reported. For example DePalma (1973) has claimed to measure a weight change in an experiment where two counter-rotating flywheels were forced to precess, and Laithwaite demonstrated that a heavy flywheel could be lifted with little effort if it were allowed to precess along a particular helical path (Davidson, 1989). Schauberger also created a precise helical path to induce water precession, and when pumped at high velocities the vortical flow exhibited a centripetal, imploding characteristic that avoided contact with the pipe walls. During such a flow he claims to have measured a negative resistance (i.e. an energy gain) and observed a peculiar bluish glow around the apparatus. Schauberger also noted that water at 4 degrees Celsius would be the most energetic. Did Schauberger discover how to induce a zero-point energy coherence with something as simple as water flow?

There is another energy anomaly associated with water that has similarities to Schauberger's observations. When water is excited by ultrasonics it emits a bluish glow called sonoluminescence (Walton and Reynolds, 1984). Barber and Putterman (1991) have experimentally shown that the photon emissions are too rapid to be from electron atomic transitions and that the phenomenon represents an energy amplification on the order of a 100 billion. Also water at 3 degrees Celsius yield greater emissions than water at room temperature (Hiller, et al., 1992). Nobel laureate Schwinger (1993) has proposed a zero-point energy interaction to explain the anomaly.

Nuclei Motion in Solid State

A zero-point energy coherence might also be triggered in highly stressed, polarized dielectrics. T.T. Brown (Sigma, 1977) discovered that a charged capacitor exhibited a unidirectional thrust in the direction of the positive plate. The greater the permittivity of the dielectric, the greater the thrust. The largest surge would occur at the moment of dielectric breakdown. Brown attempted to prove that the source of thrust was due to an electrogravitic interaction with experiments in vacuum and under oil in order to eliminate ion propulsion. During dielectric breakdown the lattice nuclei abruptly move in response to a sudden, disruptive shock wave from the plasma discharge surging through the material. The abrupt event might produce a large enough ZPE coherence to manifest a gravitational effect.

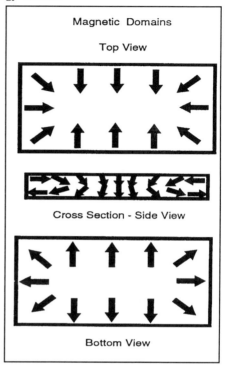

Figure 3. Magnetic domain alignment in barium ferrite after conditioning.

Another energetic anomaly that occurs in solids is fractoemission. When a crystal is cracked, it can sometimes manifest a persistent plasmoid-like form that is akin to earthquake lights (ball lightning formation from ground fissures). Preparata (1991) has noted fractoemission can yield highly accelerated electrons as well as light emissions which can sometimes persist for hours after the original fracturing event. At the crack boundary nuclei motion could trigger a coherence in the ZPE which becomes the energy source for maintaining the plasmoid. Similarly, Shoulders (1991) has discovered a micron size, charged plasmoid which can persist indefinitely on a dielectric surface that he calls "electrum validum" (EV). He creates the EV with a high voltage pulsed discharge from a mercury tip electrode and guides it via electric fields along micron size groves etched in a dielectric material. When the EV

Vacuum Energy Vortices

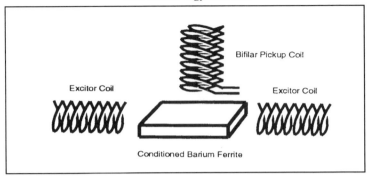

Figure 4. Placement of coils in Sweet's Vacuum Triode Amplifier.

hits a conductor it yields a discharge whose energy is greater than the original pulse used in its creation. Puthoff (1990) suggests that the energy is from the ZPE.

Lattice Twisting

Another method that might cohere the ZPE is to induce a ratcheting, semi-vortical motion of nuclei by abruptly twisting a crystalline lattice. Lattice twisting can occur in ferromagnetic materials when subject to alternating magnetic fields. As the magnetic domains shift, they can launch acoustical spinor waves (Cieplak, 1980). An abrupt lattice twist can occur if the magnetic material is driven to hysteresis saturation, and then pulsed oppositely. The saturation state elastically stresses the lattice, and the reverse pulse triggers the lattice to snap back. Aspden (1990) has identified hysteresis saturation as a significant state for generating energetic anomalies with magnetic materials. The nuclei motion from the lattice twist can launch the vacuum energy vortex which manifests as an excess magnetic pulse. This type of activity could be occurring in the stators of Adam's (1993) pulsed magnetic motor, an invention claimed to produce excessive power.

Sweet (1991) also appears to utilize lattice twisting in the conditioned barium ferrite magnets of his solid state energy invention known as the "vacuum triode amplifier" (VTA). Normally barium ferrite is used for permanent magnets, and its domains do not readily shift. Instead, Sweet cracks and loosens the lattice itself with the conditioning process. The barium ferrite block (6x4x1 inches) should be sintered by the manufacturer such that the ceramic is not overly hard. The conditioning is similar to how

58

manufacturers make permanent magnets: An a.c. current is impressed on a coil surrounding the material to erase any residual magnetization. Then a large pulse from a capacitor bank (a typical manufacturer uses 100 microfarads at 15KV) is fired through the coil to align the domains into a permanent magnet. Sweet's conditioning coil surrounds the (6x4) perimeter of the barium ferrite block and consists of 600 turns of No. 28 wire. He drives it at 60 Hz with a few amps and then switches a large pulse from a 6500 microfarad capacitor at 450 volts (values reported by Watson, 1993) through the coil, timed at the peak of the 60 Hz sine wave. Unless the ceramic is loosely sintered, it is unlikely one firing will crack the lattice. The barium ferrite block should then be turned over (or the coil polarity reversed) and the process repeated such that the domains are driven to the opposite polarity from the next capacitor pulse. The conditioning process should be repeated over and over, altering the polarity each time. The process is analogous to cold working a strip of metal by bending it back and forth until it breaks. The lattice will form micro cracks and loosen such that the magnetic domains appear to readily oscillate when excited by a weak a.c. magnetic field. It is really not the magnetic domains that are shifting; it would be more accurate to describe the cracked portions as acoustical domains since it is the lattice grains that are shifting. Thus in a straight forward manner Sweet has created a twistable solid state lattice that exhibits an acoustical resonance at the conditioning frequency (60 Hz).

Within the micro cracks of the conditioned barium ferrite apparently occurs the phenomenon that coheres the ZPE. Perhaps the shifting lattice's grains induce fractoemission in the boundaries between them. At the crack boundaries nuclei motion could be triggering a coherence in the ZPE which maintains the fractoemission. If grains twisting against each other induce fractoemission within the interior of the conditioned barium ferrite, a coherent plasma would be embedded within a solid, and this plasma could to be directly coupled to the zeropoint energy. The embedded plasma is controlled by the action of the twisting domains. The plasma in the boundary

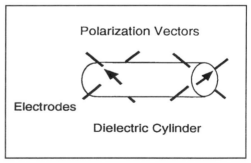

Figure 5. Flexing polarization vectors in an ultrasonic transducer creates spinor waves.

between two oppositely twisting grains would be subject to dual counter-rotating vortical stimulation. The conditioned barium ferrite has the magnetic domains on the top surface aligned oppositely to those on the bottom surface, especially near the edge (Figure 3). The domain alignment follows the flux lines of the conditioning coil. Near the edges of the top and bottom faces, the domains are aligned nearly flat. (Sweet demonstrates this by placing a thin steel strip edgewise on the ceramic face. In the center it stands vertically; it tilts more horizontally as it is placed closer to the edge.) When excited by an a.c. magnetic field from the side, the top and bottom domains will twist in opposite directions as they oscillate. In the micro cracks between these oppositely twisting domains, pair production of fractoemission plasmoids exhibiting opposite helicity might occur. QED requires that vortical coherent forms arising from the ZPE occur in pairs to conserve angular momentum. If such plasmoid pairs are generated throughout the interior of the ceramic, they could integrate into two macroscopic, counter-rotating, displacement current vortices.

Dual vortex action is required in order to induce a current on a series wound, bifilar coil (i.e. the windings are shorted on one end). Sweet places the bifilar coil on the face of the conditioned barium ferrite (which is driven by an a.c. magnetic field from two separate, in phase, standard wound, "excitor" coils directed at the 4 inch sides, Figure 4). In a series wound bifilar coil, current must flow up one winding while it simultaneously flows down the other. No form of standard magnetic induction could induce this type of current flow. Sweet is able to tap appreciable electric power (500 watts) from the bifilar coil, and in addition, the current from this coil is "cold", i.e. thin wires (No. 30) can conduct the power without being heated. The current does not appear to be standard electron conduction; instead the coil seems to be guiding vacuum energy displacement currents (King, 1984). Other inventors have observed this cold current effect; e.g. Moray (1978) used No. 30 wire in his 50 kilowatt radiant energy device, Bedini observed it from his "gravity field generator" (Bearden, Herold, Mueller, 1985), and Panici observed it in his battery pulsing experiments (King, 1993). To make the VTA free running, Sweet taps off some of the output power from the bifilar coil and feeds it back to the excitor coils. If too much power is fed back, it overdrives the lattice twisting, and the ceramic pulverizes. The counter rotating vortices appear to significantly cohere the ZPE.

There is another indication that the ZPE is the energy source for the VTA. When running, it loses weight. Wheeler's (1962) geometrodynamics shows there is an intimate connection between the ZPE, gravity and curving the space-time metric. Likewise, Puthoff (1989) suggests that the basis of gravity is the ZPE itself. When driven to output a kilowatt, the VTA has exhibited a weight change on the order of a pound. The only energy great enough to locally alter gravity would apparently be the ZPE.

The counter rotating vortex hypothesis may also explain another well witnessed, free running, energy machine: the Swiss ML Converter (Matthey, 1985). Two large, counter rotating, acrylic disks (similar to a Whimshurst machine) induce a colorful, swirling plasma between them. This plasma would then be the analog to the fractoemission plasma between the lattice domains in the VTA ceramic. The friction action of the spinning disks create the counter rotating plasma which induces the dual vortex, ZPE displacement currents. The principle of using counter rotating plasmas might form the basis for many future ZPE machines.

The counter rotating vortex idea may elucidate how to stimulate the core material of a caduceus coil in order to output excess energy. The caduceus coil consists of two perfectly symmetric, insulated wire coils of opposite helicity wound on a cylinder. The coil wires must crisscross each other in order to be identically symmetrical. Smith wound his coils on ferrite cores (Burridge, 1979) and Van Tassel wound his on quartz (Dollard, 1988). Bearden (1986) and King (1989) have suggested pulsing these coils to create abruptly bucking electromagnetic fields that would result in a scalar ZPE coherence. On the other hand, to receive energy from the coil would require inducing the appropriate lattice twisting on the core material. For ferrites, a conditioning process much like Sweet uses for his barium ferrite would be appropriate. The domains would have to align along the diameter of the cylinder (right angles to the cylinder's axis) with alternate directions in regions at different heights along the cylinder. The domains would shift in response to excitor coils at right angles to the caduceus coil. If the ferrite lattice has cracks in between the alternate domain regions, then lattice twisting could induce the fractoemission plasma and the behavior of the cadu-

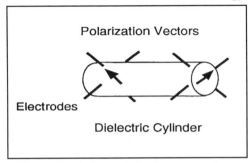

Figure 5. Flexing polarization vectors in an ultrasonic transducer creates spinor waves.

ceus coil would mimic the VTA. (It would effectively be the same as the VTA except for the shape of the barium ferrite).

Lattice twisting could also be induced on high permittivity dielectrics (barium titanate, lead zirconate, etc.) The polarization vector of the dielectric would follow the high voltage excitation from electrodes placed on the circumference of the cylinder (Figure 5). At a different position along the cylinder another set of electrodes would drive the polarization vector opposite to the previous layer in order to induce the counter twist to the lattice. Cracks between the lattice regions would offer opportunities for fractoemission. An electret whose polarization vectors are permanently aligned would be the analog to permanent magnetic material and would allow maximal lattice twisting. In ultrasonic transducer materials, the stimulation could induce counter rotating, acoustical spinor waves (King, 1993). If instead of a cylinder, a toroid is used, the spinor waves can circulate and close into standing waves. If concomitant ZPE displacement currents align with these counter rotating standing waves, then the positive feedback of such coherence might induce a sufficiently great ZPE interaction to produce a large gravitational effect.

Summary

The observed anomalies associated with ion motion in plasmas, liquids or solids along with the zero-point energy theories manifest a common theme that may be summarized into four principles:

1) The abrupt, synchronous motion of nuclei or ions cohere the ZPE.

2) Vortex motion of the ions produce even a greater effect, and there might be an optimal vortex shape around which a ZPE coherence would naturally form.

3) Higher order rotations, i.e. precession, further augment the ZPE in-

teraction.

4) A large macroscopic ZPE coherence would involve pairs of counter-rotating vortical forms since this conserves angular momentum.

Today, two free running energy inventions appear to utilize these principles and consequently produce a large energy output. The Swiss ML converter clearly has two counter-rotating plasma vortices since the corona is readily observed between the oppositely rotating acrylic disks. Unfortunately, there is not enough technical information available to allow widespread replication. On the basis of simplicity, Sweet may have achieved the most elegant energy device ever invented, yet its behavior is complex since it appears to contain multiple pairs of microscopic, counter-rotating, fractoemission plasmoids within a single ceramic block. If Sweet's invention could be replicated by the scientific community, our science and technology would enter a new era.

References

Adams, R.(Jan 1993), "The Adams Pulsed Electric Motor Generator," Nexus, pages 31-36.

Aspden, H.(1990), "Switched Reluctance Motor with Full Accommutation," U.S. Patent No. 4,975,608; ... (1993), "The World's Energy Future," Proc. Int. Sym. on New Energy, pages 1-19.

Alexandersson, O.(1990), Living Water: Viktor Schauberger and the Secrets of Natural Energy, Gateway Books, Bath, UK. Also Frokjaer-Jensen, B.(1981), "The Scandinavian Research Organization and the Implosion Theory (Viktor Schauberger)," Proc. First International Symposium on Nonconventional Energy Technology, Toronto, pages 78-96.

Barber, B.P. and S.J. Putterman (1991), "Observation of synchronous picosecond sonoluminescence," Nature 353, pages 318-320; ... (1992), "Light Scattering Measurements of the Repetitive Supersonic Implosion of a Sonoluminescing Bubble," Phys. Rev. Lett. 69, pages 3839-42.

Bearden, T. and T. Herold, E. Mueller (1985), "Gravity Field Generator Manufactured by John Bedini," Tesla Book Co., Greenville TX.

Bearden, T.(1986), Fer-De-Lance: A Briefing on Soviet Scalar Electromagnetic Weapons, Tesla Book Co., Millbrae, CA., pages 107-108.

Vacuum Energy Vortices

Bostick, W.H.(1957), "Experimental Study of Plasmoids," Phys. Rev. 106(3), page 404.

Bostick, W.H.(1966), "Pair Production of Plasma Vortices," Phys. Fluids 9, pages 2078-80.

Boyer, T.H.(1975), "Random Electrodynamics: The theory of classical electrodynamics with classical electromagnetic zero-point radiation," Phys. Rev. D 11(4), pages 790-808; ... (1969), "Derivation of Blackbody Radiation Spectrum without Quantum Assumptions," Phys. Rev. 182(5), pages 1374-83.

Boyer, T.H. (1976), "Equilibrium of random classical electromagnetic radiation in the presence of a nonrelativistic nonlinear electric dipole oscillator," Phys. Rev. D 13(10), pages 2832-45.

Brown, P.M.(1989), "Apparatus for Direct Conversion of Radioactive Decay Energy to Electrical Energy," U.S. Patent No. 4,835,433; ... (1987), "The Moray Device and the Hubbard Coil were Nuclear Batteries," Magnets 2(3), pages 6-12; ... (1990), "The Resonant Nuclear Battery," International Tesla Symposium, Colorado Springs.

Burridge, G.(1979), "The Smith Coil," Psychic Observer 35(5), pages 410-416.

Celenza, L.S. and V.K. Mishra, C.M. Shakin, K.F. Liu (1986), "Exotic States in QED," Phys. Rev. Lett. 57(1), page 55; Caldi, D.G. and A. Chodos (1987), "Narrow e^+e^- peaks in heavy-ion collisions and a possible new phase of QED," Phys. Rev. D 36(9), page 2876; Jack Ng, Y. and Y. Kikuchi (1987), "Narrow e^+e^- peaks in heavy-ion collisions as possible evidence of a confining phase of QED," Phys. Rev. D 36(9), page 2880; Celenza,L.S. and C.R. Ji, C.M. Shakin (1987), "Nontopological solitons in strongly coupled QED," Phys. Rev. D 36(7), pages 2144-48.

Cieplak, M. and L.A. Turski (1980), "Magnetic solitons and elastic kink-like excitations in compressible Heisenberg chain," J. Phys. C: Solid State Physics 13, pages L 777-780.

Cole, D.C. and H.E. Puthoff (1993), "Extracting energy and heat from the vacuum," Phys. Rev. E 48(2), pages 1562-65.

Davidson, J.(1989), The Secret of the Creative Vacuum, C.W. Daniel Co. Ltd., Essex, UK, pages 258-262.

DePalma, B.E. and C.E. Edwards (1973), "The Force Machine Experiments," privately published.

Dollard, E.(1988), "Van Tassel's Caduceus Coils," private communication. Van Tassel experimented with numerous caduceus coils that often contained quartz cores. His notes stated that the cross-over angle for the two opposing windings should be 22.5 degrees.

Dufour, J.(1993), "Cold Fusion by Sparking in Hydrogen Isotopes," Fusion Technology 24, pages 205-228.

Egely, G.(1986), "Energy Transfer Problems of Ball Lightning," Central Research Institute for Physics, Budapest, Hungary.

Forward, R.L.(1984), "Extracting electrical energy from the vacuum by cohesion of charged foliated conductors," Phys. Rev. B 30(4), pages 1700-2.

Haisch, B. and A. Rueda, H.E. Puthoff (1994), "Inertia as a zero-point field Lorentz force," Phys. Rev. A 49(2), pages 678-694.

Hathaway, G.(1991), "Zero-Point Energy: A New Prime Mover? Engineering Requirements for Energy Production & Propulsion from Vacuum Fluctuations," Proc. 26th IECEC vol. 4, pages 376-381.

Hiller, R. and S. Putterman, B. Barber (1992), Phys. Rev. Lett. 69, pages 1182-84.

Honig, W.M.(1986), The Quantum and Beyond, Philosophical Library, NY; ... (1974), "A Minimum Photon Rest Mass using Planck's Constant and Discontinuous Electromagnetic Waves," Found. Phys. 4(3), pages 367-380.

Huntley, H.E. (1970), The Divine Proportion, Dover Publications, NY.

Jandel, M.(1990), "Cold Fusion in a Confining Phase of Quantum Electrodynamics," Fusion Tech. 17, pages 493-499.

Jennison, R.C.(1978), "Relativistic Phase-Locked Cavities as Particle Models," J. Phys. A Math Gen. VII(8), pages 1525-33; ... (1989), "A New Classical Relativistic Model of the Electron," Phys. Lett. A 141(8/9), pages 377-382.

Jennison,R.C.(1990), "Relativistic Phase-Locked Cavity Model of Ball Lightning," Electronics Laboratory, University of Kent, U.K.

Johnson, P.O.(1965), "Ball Lightning and Self Containing Electromagnetic Fields," Am. J. Phys. 33, page 119.

Kalinin, Yu G. et al.(1970), "Observation of Plasma Noise During Turbulent Heating," Sov. Phys. Dokl. 14(11), page 1074; Iguchi, H.(1978), "Initial State of Turbulent Heating of Plasmas," J. Phys. Soc. Jpn. 45(4), page 1364; Hirose, A.(1974), "Fluctuation Measurements in a Toroidal Turbulent Heating Device," Phys. Can. 29(24), page 14.

Vacuum Energy Vortices

King, M.B.(1984), "Macroscopic Vacuum Polarization," Proc. Tesla Centennial Symposium, International Tesla Society, Colorado Springs, pages 99-107. Also (1989), pages 57-75.

King, M.B.(1989), Tapping the Zero-Point Energy, Paraclete Publishing, Provo, UT; ... (1991), "Tapping the Zero-Point Energy as an Energy Source," Proc. 26th IECEC vol.4, pages 364-369; ... (1993), "Fundamentals of a Zero-Point Energy Technology," Proc. Int. Sym. on New Energy, pages 201-217.

Kiwamoto, Y. and H. Kuwahara, H. Tanaca (1979), "Anomalous Resistivity of a Turbulent Plasma in a Strong Electric Field," J. Plasma Phys. 21(3), page 475.

La Violette, P.A.(1985), "An introduction to subquantum kinetics...," Intl. J. Gen. Sys. 11, pages 281-345; ... (1991), "Subquantum Kinetics: Exploring the Crack in the First Law," Proc. 26th IECEC vol. 4, pages 352-357.

Mallove, E.F.(1992), "Protocols for Conducting Light Water Excess Energy Experiments," Fusion Facts 3(8), page 15; Noninski, V.C.(1992), "Excess Heat during the Electrolysis of a Light Water Solution of K_2CO_3 with a Nickel Cathode," Fusion Tech. 21, pages 163-167.

Matthey, P.H. (1985), "The Swiss ML Converter - A Masterpiece of Craftsmanship and Electronic Engineering," in H.A. Nieper (ed.), Revolution in Technology, Medicine and Society, MIT Verlag, Odenburg.

Michrowski, A.(1993), "Vacuum Energy Developments," Proc. Int. Sym. on New Energy, pages 407-417.

Mills, R.L. and S.P. Kneizys (1991), "Excess Heat Production by the Electrolysis of an Aqueous Potassium Carbonate Electrolyte and the Implications for Cold Fusion," Fusion Tech. 20, pages 65-81.

Milonni, P.W. and R.J. Cook, M.E. Goggin (1988), "Radiation pressure from the vacuum: Physical interpretation of the Casimir force," Phys. Rev. A 38(3), pages 1621-23.

Moray, T.H. and J.E. Moray (1978), The Sea of Energy, Cosray Research Institute, Salt Lake City.

Mizuno, T. and M. Enyo, T. Akimoto, K. Azumi (1993), "Anomalous Heat Evolution from $SrCeO_3$ - Type Proton Conductors during Absorption/Desorption of Deuterium in Alternate Electric Field," 4th Int. Conf. on Cold Fusion. Abstract in Fusion Facts, Dec. 1993, page 30.

Vacuum Energy Vortices

Pappas, P.T.(1991), "Energy Creation in Electrical Sparks and Discharges: Theory and Direct Experimental Evidence," Proc. 26th IECEC vol. 4, pages 416-423.

Preparata, G.(1991), "A New Look at Solid-State Fractures, Particle Emission and Cold Nuclear Fusion," Il Nuovo Cimento 104 A(8), page 1259; ... (1990), "Quantum field theory of superradiance," in Cherubini, R., P. Dal Piaz, B. Minetti (editors), Problems of Fundamental Modern Physics, World Scientific, Singapore.

Prigogine, I. and G. Nicolis (1977), Self Organization in Nonequilibrium Systems, Wiley, NY; Prigogine, I. and I. Stengers (1984), Order Out of Chaos, Bantam Books, NY.

Puthoff, H.E.(1987), "Ground state of hydrogen as a zero-point fluctuation determined state," Phys. Rev. D 35(10), pages 3266-69.

Puthoff, H.E.(1989), "Gravity as a zero-point fluctuation force," Phys. Rev. A 39(5), pages 2333-42.

Puthoff, H.E.(1990), "The energetic vacuum: implications for energy research," Spec. Sci. Tech. 13(4), pages 247-257.

Rausher, E.A.(1968), "Electron Interactions and Quantum Plasma Physics," J. Plasma Phys. 2(4), page 517.

Reed, D. (1992), "Toward a Structural Model for the Fundamental Electrodynamic Fields of Nature," Extraordinary Science IV(2), pages 22-33; ... (1993), "Evidence for the Screw Electromagnetic Field in Macro and Microscopic Reality," Proc. Int. Sym. on New Energy , pages 497-510; ... (1994), "Beltrami Topology as Archetypal Vortex," Proc. Int. Sym. on New Energy, (in press).

Reinhardt, J. and B. Muller, W. Greiner (1980), "Quantum Electrodynamics of Strong Fields in Heavy Ion Collisions," Prog. Part. and Nucl. Phys. 4, page 503.

Samokhin, A.(1990), "Vacuum energy - a breakthrough?" Spec. Sci. Tech. 13(4), page 273.

Scheck, F.(1983), Leptons, Hadrons and Nuclei, North Holland Physics Publ., NY, pages 213-223.

Schwinger, J.(1993), "Casimir light: The source," Proc. Natl. Acad. Sci. USA 90, pages 2105-6.

Scully, M.O. and M. Sargent (March 1972), "The Concept of the Photon," Physics Today, page 38.

Vacuum Energy Vortices

Senitzky, I.R.(1973), "Radiation Reaction and Vacuum Field Effects in Heisenberg-Picture Quantum Electrodynamics," Phys. Rev. Lett. 31(15), page 955.

Sethian, J.D. and D.A. Hammer, C.B. Whaston (1978), "Anomalous Electron-Ion Energy Transfer in a Relativistic-Electron-Beam Heated Plasma," Phys. Rev. Lett. 40(7), page 451; Robertson, S. and A. Fisher, C.W. Roberson (1980), "Electron Beam Heating of a Mirror Confined Plasma," Phys. Fluids 32(2), page 318; Tanaka, M. and Y. Kawai (1979), "Electron Heating by Ion Acoustic Turbulence in Plasmas," J. Phys. Soc. Jpn. 47(1), page 294.

Shoulders, K.R.(1991), "Energy Conversion Using High Charge Density," U.S. Patent No. 5,018,180.

Sigma, R. (1977), Ether Technology, Tesla Book Co., Millbrae CA.

Singer, S.(1971), The Nature of Ball Lightning, Plenum Press, NY; Silberg, P.A.(1962), "Ball Lightning and Plasmoids," J. Geophys. Res. 67(12), page 4941.

Spence G.M.(1988), "Energy Conversion System," U.S. Patent No. 4,772,816.

Storms, E.(1991), "Review of Experimental Observations about the Cold Fusion Effect," Fusion Tech. 20, pages 433-477.

Suzuki, M.(1984), "Fluctuation and Formation of Macroscopic Order in Nonequilibrium Systems," Prog. Theor. Phys. Suppl. 79, pages 125-140; Hasegawa, A.(1985), "Self-Organization Processes in Continuous Media," Adv. Phys. 34(1), pages 1-41; Firrao, S.(1984), "Physical Foundations of Self-Organizing Systems Theory," Cybernetica 17(2), pages 107-124; Haken, H.(1971), Synergetics, Springer Verlag, NY.

Sweet, F. and T.E. Bearden (1991), "Utilizing Scalar Electromagnetics to Tap Vacuum Energy," Proc. 26th IECEC vol. 4, pages 370-375; ... (1988), "Nothing is Something: The Theory and Operation of a Phase-Conjugate Vacuum Triode;" ... (1989), private communication.

Walton, A.J. and G.T. Reynolds (1984), "Sonoluminescence," Adv. Phys. 33(6), pages 595-600.

Watson, M. (1993), Watson, D. (1994) private communication. Many investigators have attempted using a capacitance of 6500 microfarads at 450 volts without success. A greater energy pulse or a greater a.c. current may be necessary to fracture the ceramic. If the lattice grains cannot be loosened via magnetic stimulation, then other metallurgic means should be explored as a preconditioning process. The values given are sufficient to align the magnetic domains in barium ferrite, and can be applied once the ceramic has pockets of loose, freely moving grains.

Vacuum Energy Vortices

D. Watson reported successful conditioning of barium ferrite by subjecting the ceramic to an electrostatic field between two conductive plates held at 20 KV D.C. for about 20 minutes before the magnetic conditioning step. Barium ferrite is an insulator (unlike strontium ferrite), and such an electrostatic field could stretch the lattice. Also, two separate coils around the 6x4 perimeter were used: a standard coil for the capacitor pulse and a 400 turn, bifilar coil for the 60 Hz entrainment signal. The bifilar coil was driven in the bucking configuration, and thus it would not induce magnetic domain motion. Perhaps the electrostatic conditioning excites interstitial plasmoids within the lattice and these respond to the scalar excitation of the bucking fields.

Wheeler, J.A.(1962), Geometrodynamics, Academic Press, NY.

Winterberg, F.(1990), "Maxwell's Equations and Einstein-Gravity in the Planck Aether Model of a Unified Field Theory," Z. Naturforsch. 45 a, pages 1102-16; ... (1991), "Substratum Interpretation of the Quark-Lepton Symmetries in the Planck Aether Model of a Unified Field Theory," Z. Naturforsch. 46 a, pages 551-559.

Vacuum Energy Vortices

The Super Tube

April 1996

Abstract

The vacuum polarization of atomic nuclei offer a means to cohere the zero-point energy (ZPE) when plasma ions undergo abrupt surges. Examples include the ion-acoustic resonance and the abnormal glow discharge. A "super tube" which combines the use of the hollow cathode discharge, radioactive cathodes, and inert gas mixtures should create an excessive energy output when tuned to cycle in the abnormal glow discharge regime. Output energy in the form of large voltage spikes are efficiently absorbed by a Pulse Current Multiplier circuit which might offer a solid state means of tapping the zero-point energy.

Introduction

There is an interesting dichotomy in theoretical physics today. The equations of quantum electrodynamics utilize an all-pervading energy inherent to the fabric of space called the zero-point energy (ZPE). The energy takes the form of random fluctuations of high density electric fields which can give rise to short lived pairs of elementary particles that are described as virtual. There is also a school of thought represented by a wide range of published papers called random electrodynamics (Boyer, 1975) that describe a real zero-point energy spectrum which interacts with the elementary particles to manifest quantum behavior. The theories show how the ZPE is the basis of gravity (Puthoff, 1989) and inertia (Haisch, 1994), and how it can be an energy source without violating thermodynamics (Cole, 1993, King, 1989). Yet the majority of physicists believe the energy is not real, but rather just a fiction of their mathematics. Note that the zero-point energy terms are quite necessary to accurately predict phenomena in quantum mechanics. Kaku (1995) describes the situation as the greatest disparity in the history of science: The equations describe an energy density on the order of 10^{100} ergs per cubic centimeter, yet most scientists believe there is actually zero.

What could be the reason for this disparity? Perhaps the paradigm shift is just too big. The implications of accepting a real zero-point energy drastically change our concept of reality. For example, at large energy densities, space-time pinches into a blackhole. Wheeler's (1962) theory of geometrodynamics describe this phenomena where space pinches into wormholes on the scale of the Planck length (10^{-33} cm) that channel the zero-point energy flux through a physical fourth dimension which can potentially connect to other, parallel, three dimensional universes (Wolf, 1990). Not only does such a description support Everett's (1957) many world's interpretation of quantum mechanics, but it also might provide a physical foundation for hyperspatial theories of leading physicists such as Hawkings and Coleman (see Kaku, 1995). Wheeler describes, as this flux passes through our universe it manifests a plethora of mini whiteholes (flux entering) and mini blackholes (flux leaving) in a chaotic state of turbulence called the "quantum foam." A temporary vorticity in this flow would manifest as a virtual elementary particle. A stable vorticity on the scale of the Planck length might be the physical basis of strings supporting superstring theories (Peat, 1988). A real zero-point energy implies the existence of physically real higher dimensions of space.

If higher dimensions exist, then why can't we see them? Ralphs (1994) answers the question quite nicely from the standpoint of perception psychology: the human brain is simply not wired to perceive beyond three physical dimensions. Our concept of space and time is simply a notion filtered and limited by human perception. Perhaps the ancient Hindu and Buddhist philosophies are correct: our concept of the everyday world is really "maya" or illusion. Could the Eastern philosophies actually be right and the Western view of "bottom line" reality be wrong? These implications are probably a little too much for a pragmatic Western scientist to swallow.

A physicist who accepts the existence of the zero-point energy might argue that for all practical purposes the energy available for use is effectively zero because its action is random and chaotic. Does not the second law of thermodynamics imply random things must forever remain random? This is generally true for closed systems in thermodynamic equilibrium, but the Nobel prize winning work of Prigogine (1977) describe other systems that may evolve from chaos to order if the system fulfills the appropriate conditions: 1) It is nonlinear, 2) far from equilibrium, and 3) have an energy flux through it. The behavior of the zero-point energy, especially in its interaction with plasmas, fulfills these conditions and opens the possibility for tapping it as an energy source (King, 1989).

Perhaps the most relevant objection by the scientific community to the notion of tapping energy from space is, where is the experiment? Any hypothesis, especially one that violates the current paradigm, requires an experiment that can be readily repeated by the scientific community to validate it. This paper will review some current plasma research which manifests excessive energy anomalies, and combine the results with the ideas of a number of inventors to propose the "super tube," a plasma tube that optimally coheres the zero-point energy. Also described will be a circuit that efficiently absorbs high voltage spikes, which might provide the foundation for a solid state method of tapping the zero-point energy. It is an experiment which can detect and cohere the ZPE for a net energy output while self running that will best convince a skeptical scientific community.

Ions and Vacuum Polarization

Why can't ordinary electric circuits and antennas detect the ZPE? Quantum electrodynamics answers the question by describing the interaction (i.e. vacuum polarization) of the elementary particles with the ZPE. There is a gradual increase in vacuum polarization intensity approaching closer to the particle. Senitzky (1973) concludes the vacuum fluctuations and the elementary particle are actually inseparably intertwined; i.e., there are no hard boundaries where the particle ends and the vacuum polarization begins. In Wheeler's view the hyperspatial ZPE flux maintains the particle's existence much as a flowing stream maintains an eddy. Also, the different elementary particles have a different vacuum polarization characteristic depending on their situation (Scheck, 1983, Reinhardt, 1980). Electrons, especially bound or in the conduction band, are described as a "smeared cloud" in which the zero-point fluctuations are effectively in thermodynamic equilibrium. In this view the bound electron would literally be a cloud of net negatively polarized vacuum fluctuations and not a point particle. Puthoff (1987) shows how the zero-point fluctuations and the bound electron reach an equilibrium to produce the stable ground state for hydrogen. The smeared cloud of conduction band electrons offer no concentrated vacuum polarization, or far from equilibrium interaction, that could induce a self-organization in the zero-point fluctuations. Thus typical electric circuits and antennas are not able to cohere the ZPE into an energy source.

On the other hand, atomic nuclei exhibit a concentrated vacuum polarization characteristic with the polarization lines converging steeply and stably toward the nuclei center. This offers opportunities for the nuclei to create an augmenting vacuum energy interaction whenever they undergo abrupt motion. Exotic coherent vacuum states are observed in accelerator experiments involving colliding heavy ions (Celenza, 1986, et al.). In the hyperspatial flux model, a surge in nuclei motion bend the ZPE flux, aligning more of it into our three space. During the surge a greater vacuum polarization would occur manifesting as an excessive transient in potential or voltage. High frequency voltage spikes (Kalinin, 1970, et al.), runaway electrons (Kiwamoto, 1979) and anomalous heating (Sethian, 1978, et al.) are observed in plasmas during ion-acoustic resonance where the plasma nuclei are oscillating. Few researchers have been looking for excessive energy anomalies in academic plasma research. However, a notable exception is Chernetskii from the University of Moscow who claimed to produce an over unity energy gain from the ZPE in his plasma discharge experiments (Samokin, 1990, Michrowski, 1993). Also Dufour (1993) has measured anomalous heat in his simple sparking experiments. There is also coherent ion motion occurring in sonoluminescence (Barber, 1991) where ultrasonic excitation of water produces an anomalous bluish glow. Nobel laureate Schwinger (1993) suggests the source of the anomalous energy is the ZPE. Also there is excess heat and coherent proton motion in the light water, "cold fusion" experiments where a nickel (or palladium) hydride has (to get the best results) a pure crystalline structure that is fully loaded and subject to pulsed, electrical stimulation (Patterson, 1996, Piantelli, 1995). There is ample experimental evidence suggesting that abrupt, synchronous motion of nuclei in plasmas can induce a zero-point energy coherence.

The principle of using coherent plasma ion motion was the basis of the radiant energy invention of T. Henry Moray (1978, also Resines, 1989, Sego, 1981, King, 1989). Moray stressed the importance of maintaining ion oscillations in his plasma tubes, which used radioactive cathodes as an ionization source. In the 1920's Moray created a two tube machine that was able to free run and light a single bulb. By the late 1930's he created multiple stages using 30 tubes that output 50 kilowatts of electricity. Each stage oscillated at a frequency higher than the subsequent stage, and the output of each was rectified onto capacitors to be the subsequent stage's input. Moray's radiant energy invention was well witnessed and perhaps the best documented "free energy" machine in history. Many researchers have investigated plasma tubes, occasionally observing some energetic transients, but found it difficult to maintain steady state ion oscillations. One problem is that the frequency of the ion-acoustic resonance (a function of the ion density) increases during the energetic event which causes more ionization - a highly nonlinear situation. Also, the standard, collisional, electron modes of the plasma tend to exhibit heating losses, and remove energy from the electrical modes. Attempting to maintain ion oscillations via holding a steady-state ion-acoustic resonance has proven to be quite difficult in practice. Perhaps there is a better way to induce abrupt, synchronous ion motion.

Abnormal Glow Discharge

Correa (1995) has recently discovered a method to induce stable plasma ion pulsing in his tubes, and moreover, has demonstrated an over unity electrical energy gain in his experiments. The method relies on carefully tuning the tubes to operate in the abnormal glow discharge regime. The abnormal glow discharge is a glowing, cold plasma located just above the cathode that occurs just before a vacuum arc discharge (spark) whenever the tube is charged slowly. It is characterized by a voltage buildup (Figure 1) where the glow plasma polarizes yielding a large effective capacitance for the tube. When the voltage reaches a certain threshold, the plasma suddenly snaps from this highly polarized state causing abrupt ion motion and a negative resistance voltage-current characteristic. The abrupt ion surge coheres the zero-point energy. In normal tube operation the tube continues to draw current and proceeds into the high loss, vacuum arc regime. Correa's patents demonstrate how to limit the current using an appropriate capacitance in parallel to the tube to quench the arc discharge and allow recovery back to the voltage build-up phase. By appropriately adjusting the capacitor recharging circuit, Correa can tune the abnormal glow discharge cycle to a frequency of choice (anywhere from 10 Hz to 10 KHz), and the tube outputs an over unity energy gain. It appears that Correa has discovered the principle with which T. H. Moray tuned his ion oscillator tubes.

Correa's method can be further enhanced by the use of a hollow cathode which produces a more powerful version of the abnormal glow discharge called the hollow cathode discharge, or pseudospark (Gundersen, 1989). In the mid 1960's scientists in the USSR discovered that a plate cathode with a small hole backed by an attached metallic cylinder could yield a tremendously powerful arc discharge whenever the back cylinder contained a pre-excited, glowing, cold plasma. Applications of this effect focused on the abrupt switching of large currents where the switching could be triggered by a small external excitation (e.g. a light pulse in the back lighted thyratron). The descriptions

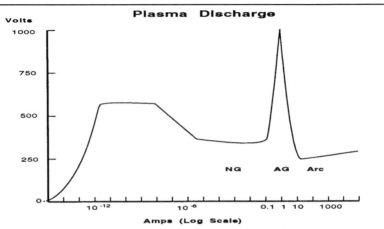

Figure 1 Voltage - current characteristic for an electric discharge from gas to plasma showing the regimes of normal glow (NG), abnormal glow (AG), and vacuum arc (Arc). Note the steep negative resistance from the peak of the abnormal glow to the beginning of the vacuum arc.

of the hollow cathode discharge note anomalies, and its behavior is similar to the abnormal glow discharge. Moreover, during the transient toward vacuum arc discharge, a portion of the hollow cathode plasma surges through the hole (Figure 2) producing a vortex ring circulation (akin to forcing smoke through a small aperture to create smoke rings). The transient appears to manifest the inception of a ball lightning plasmoid (Lewis, 1995) which could further amplify the zero-point energy interaction. A hollow cathode discharge tube tuned to cycle in the abnormal glow discharge regime forms the basis of the super tube.

Depending on the gas pressure within the tube, considerable voltage can be required to induce the glow discharge. Pre-ionizing the hollow cathode plasma greatly reduces the firing voltage. It is well known that coating a radioactive material on a cathode causes pre-ionization and avoids the need to heat the cathode (e.g. McElrath, 1936). Many inventors of free-running energy machines used this principle, e.g. Moray (1978), Papp (1984), Brown (1989). Coating the interior of the hollow cathode with radioactive material pre-ionizes the cathode plasma, augments the negative resistance, and lowers the voltage needed to operate in the abnormal glow discharge regime.

The gas mixture and pressure are likewise important and could significantly increase the energy gain of the super tube. Papp (1984) discovered tremendous energy gains by creating plasmas in a mixture of the inert gases under high pressure. Cooke (1983) has suggested that

76

the noble gases are specially energetic because their nuclei are "inter-dimensional node points" that tend to channel a hyperspatial energy flux into our three space. Cooke's ideas might gain support from a peculiar anomaly associated with sonoluminescence: Mixing inert gases with the water increases the light output 30 fold (Crum, 1995). In a normal, unenergized state there is negligible interaction between inert gas molecules. However, in the plasma state the ions tend to form into clusters. If mercury vapor is also added to the mixture, then plasmoid clusters packing anomalous energy similar to Shoulder's EV's (1991) might occur. Papp produced an explosive event from his noble gas mixture under high pressure sufficient to drive

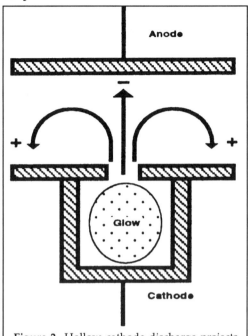

Figure 2 Hollow cathode discharge projects plasma through the aperature. The anode repulses the ions to start a vortex ring circulation.

an engine. Such mixtures could dramatically augment the output of the super tube.

A free running, two-stage device similar to T. H. Moray's first radiant energy invention can be constructed with two super tubes operating in the abnormal glow discharge regime. The output of each is rectified onto capacitors which gradually charge to become the mutual input current source for the tubes. Once the principles for the two stage device are understood, a multi stage machine can be designed.

Solid State Methods

The super tube is designed to maximize the ZPE coherence per pulse. But what if we work on a smaller scale using a large number of weak pulses that still produce excessive energy? Then the opportunity arises for solid state ZPE coherers.

A number of investigators have suggested that any abrupt electrical discharge produces anomalously excessive power. Pappas (1991) hypothesized all sparks produce excessive energy. Shoulders (1991) demonstrated that excessive energy can be created in any electrical discharge that produces EV's. What if every spark naturally produces s u b m i c r o n plasmoids (even smaller than Shoulder's EV's)?

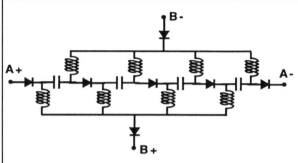

Figure 3 Pulse Current Multiplier (PCM) converts unipolar voltage spikes (input on the A terminals) to current pulses (output on the B terminals). Four stages are shown, but in practice ten or more should be used.

Finally, there is abrupt ion motion on the inception of any spark, fulfilling Moray's hypothesis. There is a strong likelihood of some ZPE coupling in any abrupt electrical discharge.

In fact some investigators have proposed that an abrupt voltage pulse with only (vacuum) displacement current and minimal charge motion is sufficient to produce anomalously excessive energy. Bearden (1993) proposed this is his "final secret, " and Hyde (1990) used this principle in designing his invention (a mechanical electrostatic field chopper) that reportedly output 20 KW while self running. Hyde stressed there should be no corona present whatsoever. The Swiss ML Converter could likewise be simply explained if indeed there were ex-

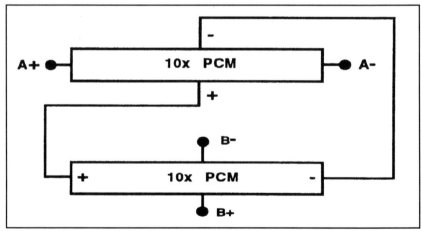

Figure 4 Combining two 10-stage Pulse Current Multipliers in series to create a 100x PCM. Input is on the A terminals; output is on the B terminals.

cessive energy in all electrostatic pulses.

Electrostatic pulses or abrupt voltage spikes are notoriously difficult to tap efficiently as an energy source. Many inventors have directed high voltage pulses to recharge batteries with claims of over unity. Unfortunately such pulsing often damages the battery, and of course, an energy machine that cannot self run, but instead relies on

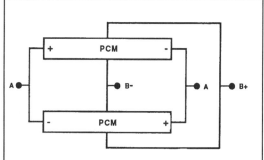

Figure 5 Bipolar Pulse Current Multiplier can convert bipolar voltage spikes (input on the A terminals) to unipolar current pulses (output on the B terminals). Two PCM's are connected in the polarity shown giving full wave rectification of voltage spikes.

batteries, is unconvincing to skeptics. Clearly what is needed is a method to convert voltage spikes to a useful form of power, preferably DC.

Hyde (1990) had to solve this problem in order to make his machine free running. His solution (described in figure 6 of his patent, also King, 1993) took advantage of the fact that even though the individual electrostatic pulses were of high voltage, the current (and thus power) of each was very low. This allowed Hyde to make a voltage divider / current multiplier circuit utilizing a large number of inexpensive capacitors and diodes that could be run in excess of their voltage specification without damage. High voltage electrostatic pulses were converted to amplified current pulses at a lower voltage which were then rectified to DC by standard means. Hyde empirically solved a difficult engineering problem, and the solution allowed him to build a self running energy machine.

An easy to build, efficient, current multiplier would facilitate widespread experiments to investigate claims of excessive energy in electrostatic pulses or sparks, as well as form the basis for a solid state zeropoint energy machine. The following Pulse Current Multiplier (PCM) was inspired by Hyde's invention. The PCM is based on the standard voltage division technique of charging a bank of capacitors in series, then discharging them in parallel. Figure 3 illustrates four stages of the PCM. In practice it is desirable to have ten or more stages. The high voltage pulse is input across the A terminals at the polarity shown. The output current pulse occurs across the B terminals. The inductors' purpose is to block the sharp input voltage pulse (via high impedance) and channel it down the (low impedance) series path which should be physi-

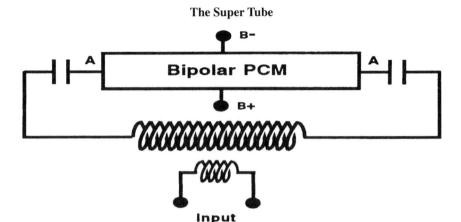

Figure 6 Tesla coil drives a bipolar PCM through air-gap capacitors to test if high voltage electrostatic pulses can generate an energy anomaly. In this control test there must be no corona.

cally short. The sharper the voltage spike, the better the performance. The inductances should be the minimum value necessary to block the input, for they must allow the output current (as a slow rise time pulse) to pass with minimal attenuation. Hyde did not show any blocking inductances in his patent. Since he developed his circuit empirically, perhaps long or coiled output wires provided a sufficient blocking inductance to achieve the same behavior. Ferrite beads might also suffice. Since the PCM uses inexpensive components, it offers an economical investigation of a possible ZPE coherence associated with high voltage pulses and discharges.

A ten stage PCM amplifies the current by a factor of ten. Two such PCM's can be combined in series to create a multiplication factor of 100 (Figure 4). The output from the first PCM is impressed across the input of the second. Since the pulses entering the second PCM will not be as sharp, larger blocking inductances will be required in it. The principle can be extended to a third PCM in series to give a multiplication factor of 1000. Thus a sharp 40 KV spike could be stepped down to a less steep 4 KV pulse and then to a wider 400 volt pulse. The second and third PCM's will require more robust circuit components since appreciable current is being accumulated. Once the pulses become sufficiently wide, then standard, solid state switching technology can be used instead of the blocking inductances to create a PCM via standard electrical engineering, voltage division techniques. Such a PCM could be employed last in the series. An appropriate configuration of PCM's can thus become an efficient voltage spike converter.

The Super Tube

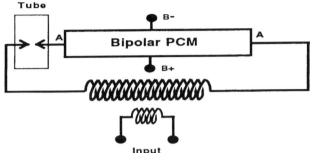

Figure 7 Tesla coil drives a spark gap or plasma tube through a bipolar PCM to convert voltage spikes to useful power. Any material that manifests abrupt ion motion in response to high voltage pulses may be substituted in place of the tube to create a solid state method of tapping the zero-point energy.

The PCM can only absorb unipolar input pulses. If both positive and negative pulses are supplied, two PCM's can be combined to create a bipolar pulse current multiplier, Figure 5. Here the input pulse naturally channels through the appropriate PCM, gated by the diodes similar to a full wave rectifier. The bipolar PCM can convert the energy of both positive and negative voltage spikes. A small, low powered, Tesla coil can be used to create high voltage pulses of dual polarity which can stimulate the bipolar PCM via electrostatic (capacitive) coupling (Figure 6) to investigate Bearden's hypothesis, or use a spark gap or tube (Figure 7) to investigate Pappas' hypothesis. Also, in place of the tube could be any material (e.g. a proton conductor, Mizuno, 1993) that manifests abrupt ion motion in response to the voltage pulses. The setup facilitates empirical testing of many materials to discover which produces the best energy gain. It is interesting to note that the slow switching diodes of the PCM fulfill Bearden's (1993) requirement of a "degenerate conductor" where conduction current is delayed so as not to drain the supplying pulsed voltage source. A Whimshurst machine could likewise provide sharp unipolar electrostatic discharges to the PCM. If excess output is observed, then the Swiss ML Converter could be explained. If there is indeed excess energy embedded in high voltage spikes or sparks, the PCM could convert this energy to a form that could be efficiently rectified to DC. The DC could drive the input pulsing source and allow the creation of an economical, solid state, self running circuit that taps the zero-point energy.

81

The Super Tube

Summary

The most important step in convincing the scientific community that energy can be extracted from the ZPE is creating an experiment in the form of a self running energy machine that can be freely replicated. There are both theoretical descriptions of nuclei vacuum polarization as well as energetic anomalies observed in plasma ion-acoustic resonances to suggest that the abrupt motion of plasma ions can yield a ZPE coherence. Correa's patents have shown how to tune plasma tubes to oscillate in the abnormal glow discharge regime to produce controlled ion surges while avoiding the losses associated with vacuum arcs. He was able to demonstrate an over unity energy gain. Correa's discovery can be further enhanced by the use of hollow cathode discharges to produce an even greater ion surge. The use of radioactive cathodes as T.H. Moray discovered pre-ionizes the glow plasma, and greatly reduces the firing voltage. The use of inert gas mixtures as Papp discovered may further augment the energy output as well. Combining all the ideas leads to the super tube, a plasma tube that optimally coheres the zero-point energy.

Since ion surges often manifest excess energy as very sharp voltage spikes, a pulse current multiplier circuit has been suggested as a means to efficiently step down large voltage spikes and convert the power to useful current pulses. PCM's can be combined in series or parallel to produce greater current multiplication or bipolar pulse rectification. The PCM can also be used in a general test setup to receive voltage spikes from sparks or materials that contain embedded free moving ions. The ions abruptly surge in response to a stimulating voltage pulse and through ZPE coupling produce an excess voltage transient. The PCM can efficiently convert the voltage transient allowing it to be rectified to DC. If sufficient output power is generated, then it can be fed back to drive the input. This could lead to an easily built, solid state, self-running device for the scientific community to replicate that would demonstrate it is possible to tap the zero-point energy as an energy source.

References

Barber, B.P. and S.J. Putterman (1991), "Observation of synchronous picosecond sonoluminescence," Nature 353, pages 318-320; ... (1992), "Light Scattering Measurements of the Repetitive Supersonic Implosion of a Sonoluminescing Bubble," Phys. Rev. Lett. 69, pages 3839-42.

Bearden, T.E.(1993), "The Final Secret of Free Energy," Proc. Int. Sym. on New Energy, pages 61-86.

Boyer, T.H.(1975), "Random Electrodynamics: The theory of classical electrodynamics with classical electromagnetic zero-point radiation," Phys. Rev. D 11(4), pages 790-808; ... (1969), "Derivation of Blackbody Radiation Spectrum without Quantum Assumptions," Phys. Rev. 182(5), pages 1374-83.

Boyer, T.H. (1976), "Equilibrium of random classical electromagnetic radiation in the presence of a nonrelativistic nonlinear electric dipole oscillator," Phys. Rev. D 13(10), pages 2832-45.

> Mathematical analysis shows that a nonlinear dipole absorbs modes of the ZPE. The author cannot believe his result, which he criticizes. Since the nonlinear interaction fulfills Prigogine's conditions for self-organization, this work could have been the first prediction published in an orthodox journal suggesting the ZPE could be an energy source.

Brown, P.M.(1989), "Apparatus for Direct Conversion of Radioactive Decay Energy to Electrical Energy," U.S. Patent No. 4,835,433; ... (1987), "The Moray Device and the Hubbard Coil were Nuclear Batteries," Magnets 2(3), pages 6-12; ... (1990), "Tesla Technology and Radioisotopic Energy Generation," Proc. 1990 International Tesla Symposium, chapter 2, pages 85-92, International Tesla Society, Colorado Springs.

> Brown created a 5 watt nuclear battery using a weak (one Curie) radioactive source, Krypton 85. Since the radioactive source could only provide at best 5 milliwatts, Brown created an anomalous self running energy device. Brown used an LC oscillator, where the radioactive material ionized a corona around the coil. If the circuit is tuned to the ion-acoustic resonance of the corona, then the ion-oscillations could couple a ZPE coherence directly to the circuit.

Celenza, L.S. and V.K. Mishra, C.M. Shakin, K.F. Liu (1986), "Exotic States in QED," Phys. Rev. Lett. 57(1), page 55; Caldi, D.G. and A. Chodos (1987), "Narrow e+e- peaks in heavy-ion collisions and a possible new phase of QED," Phys. Rev. D 36(9), page 2876; Jack Ng, Y. and Y. Kikuchi (1987), "Narrow e+e- peaks in heavy-ion collisions as possible evidence of a confining phase of QED," Phys. Rev. D 36(9), page 2880; Celenza,L.S. and C.R. Ji, C.M. Shakin (1987), "Nontopological solitons in strongly coupled QED," Phys. Rev. D 36(7), pages 2144-48.

Cole, D.C. and H.E. Puthoff (1993), "Extracting energy and heat from the vacuum," Phys. Rev. E 48(2), pages 1562-65.

The Super Tube

Cooke, M.B.(1983), Einstein Doesn't Work Here Anymore, Marcus Books, Toronto.

Presents a hyperspace theory where the noble gas nuclei constitute interdimensional node points for channeling energy.

Correa, P.N. and A.N. Correa (1995), "Electromechanical Transduction of Plasma Pulses," U.S. Patent 5,416,391. "Energy Conversion System," U.S. Patent 5,449,989.

Fundamental discovery that a plasma tube tuned to operate at the abnormal glow discharge region exhibits an over unity energy gain. The abnormal glow discharge is a glow plasma that surrounds the cathode just before a vacuum arc discharge (spark) that occurs when the tube is slowly charged with increasing voltage. The abnormal glow exhibits a negative resistance characteristic as the tube begins an arc discharge. The patents illustrate how to make an appropriate charging circuit to cycle the tube in the abnormal glow discharge regime, control the cycle frequency, avoid the losses of the vacuum arc discharge, and rectify the excess energy onto batteries.

Crum, L.A.(1995), "Bubbles Hotter than the Sun," New Scientist 1975, pages 36-40. Summarized in Fusion Facts 6(12), June 1995, page 10.

Overviews recent research in sonoluminescence including Hiller's discovery that the presence of noble gases make the luminosity increase by a factor of 30.

Dufour, J.(1993), "Cold Fusion by Sparking in Hydrogen Isotopes," Fusion Technology 24, pages 205-228.

Everett, H.(1957), "Relative State Formulation of Quantum Mechanics," Rev. Mod.Phys. 29 (3), page 457. Also 1973, B.S. Dewitt N. Graham, The Many Worlds Interpretation of Quantum Mechanics, Princeton University Press.

Everett was the first to introduce a self-consistent formulation of quantum mechanics without invoking postulates regarding the observer. The formulation yields a superspace containing an infinite number of three dimensional universes.

Farnsworth, P.T.(1939), "Cold Cathode Electron Discharge Tube," U.S. Patent 2,184,910.

Cup shaped electrodes provide a cold cathode discharge which has some similarity to the hollow cathode discharge.

Gundersen, M.A. and G. Schaefer (1990), Physics and Applications of Pseudosparks, Plenum Press, NY.

Conference proceedings studying hollow cathode discharge phenomena. If a glow discharge is created in the interior of a hollow cathode, tremendous currents may be switched and triggered by a small external signal. The phenomena generates an intense beam of ions between the electrodes. The initial stage of the discharge exhibits anomalous behavior including a negative resistance characteristic that is a powerful version of the abnormal glow discharge.

Haisch, B. and A. Rueda, H.E. Puthoff (1994), "Inertia as a zero-point field Lorentz force," Phys. Rev. A 49(2), pages 678-694.

Hyde, W.W.(1990), "Electrostatic Energy Field Power Generating System," U.S. Patent No. 4,897,592. The invention is summarized by King (1991).

Kaku, M. (1994), Hyperspace, Anchor Books Doubleday, NY.

Layman's overview to the hyperspace theories of theoretical physics includes general relativity, nuclear standard model, quantum gravity, superstrings, many-worlds, wormholes, time warps, Hawking's universal wave function and Coleman's parallel universes. Shows that unification of physics elegantly occurs via the hyperspace theories. It is noteworthy that the fundamental action in the unifying theories is occurring at the Planck length (10^{-33} cm) setting the stage for a microscopic theory of the vacuum fluctuations.

Kalinin, Yu G. et al.(1970), "Observation of Plasma Noise During Turbulent Heating," Sov. Phys. Dokl. 14(11), page 1074; Iguchi, H.(1978), "Initial State of Turbulent Heating of Plasmas," J. Phys. Soc. Jpn. 45(4), page 1364; Hirose, A.(1974), "Fluctuation Measurements in a Toroidal Turbulent Heating Device," Phys. Can. 29(24), page 14.

King, M.B.(1989), Tapping the Zero-Point Energy, Paraclete Publishing, Provo, UT; ... (1991), "Tapping the Zero-Point Energy as an Energy Source," Proc. 26th IECEC vol 4, pages 364-369; ... (1993), "Fundamentals of a Zero-Point Energy Technology," Proc. Int. Sym. on New Energy, pages 201-217; ...(1994), "Vacuum Energy Vortices," Proc. Int. Sym. on New Energy, pages 257-269.

Lewis, E.H.(1995), "Tornados and Ball Lightning," Extraordinary Science VII (4), pages 33-37; ...(March 1996), "Tornados and Tiny Plasmoid Phenomena," New Energy News 3(9), pages 18-20; ...(Feb 1994), "Some Important Kinds of Plasmoid Traces Produced by Cold Fusion Apparatus," Fusion Facts 6(8), pages 16-17.

Overview of plasmoid phenomena including tornados, ball lightning, and microscopic EV's. Includes an abundant list of references.

Matthey, P.H. (1985), "The Swiss ML Converter - A Masterpiece of Craftsmanship and Electronic Engineering," in H.A. Nieper (ed.), Revolution in Technology, Medicine and Society, MIT Verlag, Odenburg.

McElrath, H.B.(1936), "Electron Tube," U.S. Patent 2,032,545.

Patents the use of radioactive materials to create a cold cathode.

Michrowski, A.(1993), "Vacuum Energy Developments," Proc. Int. Sym. on New Energy, pages 407-417.

Mizuno, T. and M. Enyo, T. Akimoto, K. Azumi (1993), "Anomalous Heat Evolution from SrCeO$_3$ - Type Proton Conductors during Absorption/Desorption of Deuterium in Alternate Electric Field," 4th Int. Conf. on Cold Fusion. Abstract in Fusion Facts, Dec. 1993, page 30.

Moray, T.H. and J.E. Moray (1978), The Sea of Energy, Cosray Research Institute, Salt Lake City.

Papp, J.(1984), "Inert Gas Fuel, Fuel Preparation Apparatus and System for Extracting Useful Work from the Fuel," U.S. Patent No. 4,428,193; ...(1972), "Method and Means of Converting Atomic Energy into Utilizable Kinetic Energy," U.S. Patent No. 3,670,494.

Pappas, P.T.(1991), "Energy Creation in Electrical Sparks and Discharges: Theory and Direct Experimental Evidence," Proc. 26th IECEC vol. 4, pages 416-423.

Patterson, J.(1996), "System for Electrolysis," U.S. Patent No. 5,494,559.

The first U.S. "cold fusion" patent with the claim of excess energy (2000%) granted. The Patterson cell contains hundreds of sub millimeter beads made by electroplating multiple, alternating, thin layers of very pure nickel and palladium. It uses a light water electrolyte. It is considered the best of the electrolytic cold fusion type of experiments with complete repeatability and a record gain of over 1000 in one demonstration.

Peat, E.D.(1988), Superstrings and the Search for the Theory of Everything, Contemporary Books, Chicago.

Piantelli, F.(1995), "Energy generation and Generator by means of anharmonic stimulated fusion," International Patent WO 95/20816. Summarized by W. Collis, Fusion Facts 7(6), Dec 1995, pages 14-15.

Light water cold fusion experiment using pure crystalline nickel that is fully loaded. Once the cell is triggered by a vibrational stress, ultrasonic stationary waves maintain the reaction. The reaction will cease if the temperature rises sufficiently to destroy the crystalline structure of the core.

Prigogine, I. and G. Nicolis (1977), Self Organization in Nonequilibrium Systems, Wiley, NY; Prigogine, I. and I. Stengers (1984), Order Out of Chaos, Bantam Books, NY.

Puthoff, H.E.(1987), "Ground state of hydrogen as a zero-point fluctuation determined state," Phys. Rev. D 35(10), pages 3266-69.

Puthoff, H.E.(1989), "Gravity as a zero-point fluctuation force," Phys. Rev. A 39(5), pages 2333-42.

Ralphs, J.D.(1993), Exploring the Fourth Dimension, Lewellyn, St. Paul, MN.

Describes how a Euclidian fourth dimension resolves many of the paradoxes in physics as well as begin to explain paranormal phenomena. Uses perception theory from psychology to show how higher dimensions could physically exist without our awareness of them.

Reinhardt, J. and B. Muller, W. Greiner (1980), "Quantum Electrodynamics of Strong Fields in Heavy Ion Collisions," Prog. Part. and Nucl. Phys. 4, page 503.

Resines, J.(1989), "The Complex Secret of Dr. T. Henry Moray," Borderland Sciences, Garberville, CA.

Samokhin, A.(1990), "Vacuum energy - a breakthrough?" Spec. Sci. Tech. 13(4), page 273.

Scheck, F.(1983), Leptons, Hadrons and Nuclei, North Holland Physics Publ., NY, pages 213-223.

Schwinger, J.(1993), "Casimir light: The source," Proc. Natl. Acad. Sci. USA 90, pages 2105-6.

Sego, R.(1981), "The Moray Energy Device - It's Workings," privately published, Provo, UT.

Senitzky, I.R.(1973), "Radiation Reaction and Vacuum Field Effects in Heisenberg-Picture Quantum Electrodynamics," Phys. Rev. Lett. 31(15), page 955.

Shoulders, K.R.(1991), "Energy Conversion Using High Charge Density," U.S. Patent No. 5,018,180.

> Fundamental discovery on how to launch a micron size, negatively charged plasmoid called "Electrum Validum" (EV). An EV yields excess energy (over unity gain) whenever it hits the anode or travels down the axis of an hollow coil. The excess energy apparently comes from the ZPE.

Wheeler, J.A.(1962), Geometrodynamics, Academic Press, NY.

> By combining the quantum electrodynamic description of the ZPE with general relativity, the author derives the hyperspatial flux (Planck length wormhole) model of the ZPE. The model derives an energy density on the order of 10^{93} grams/cm^3 making it the most powerful of the ZPE descriptions. If true, it could not only offer a limitless energy source, but gravitaional propulsion as well.

Wolf, F.A.(1990), Parallel Universes, Simon & Shuster, NY.

> Layman's overview of the various hyperspace theories discussed in today's physics, which suggest an infinite number of three dimensional universes co-exist with our own.

References

Barber, B.P. and S.J. Putterman (1991), "Observation of synchronous picosecond sonoluminescence," Nature 353, pages 318-320; ... (1992), "Light Scattering Measurements of the Repetitive Supersonic Implosion of a Sonoluminescing Bubble," Phys. Rev. Lett. 69, pages 3839-42.

Bearden, T.E.(1993), "The Final Secret of Free Energy," Proc. Int. Sym. on New Energy, pages 61-86.

Boyer, T.H.(1975), "Random Electrodynamics: The theory of classical electrodynamics with classical electromagnetic zero-point radiation," Phys. Rev. D 11(4), pages 790-808; ... (1969), "Derivation of Blackbody Radiation Spectrum without Quantum Assumptions," Phys. Rev. 182(5), pages 1374-83.

Boyer, T.H. (1976), "Equilibrium of random classical electromagnetic radiation in the presence of a nonrelativistic nonlinear electric dipole oscillator," Phys. Rev. D 13(10), pages 2832-45.

> Mathematical analysis shows that a nonlinear dipole absorbs modes of the ZPE. The author cannot believe his result, which he criticizes. Since the nonlinear interaction fulfills Prigogine's conditions for self-organization, this work could have been the first prediction published in an orthodox journal suggesting the ZPE could be an energy source.

Brown, P.M.(1989), "Apparatus for Direct Conversion of Radioactive Decay Energy to Electrical Energy," U.S. Patent No. 4,835,433; ... (1987), "The Moray Device and the Hubbard Coil were Nuclear Batteries," Magnets 2(3), pages 6-12; ... (1990), "Tesla Technology and Radioisotopic Energy Generation," Proc. 1990 International Tesla Symposium, chapter 2, pages 85-92, International Tesla Society, Colorado Springs.

> Brown created a 5 watt nuclear battery using a weak (one Curie) radioactive source, Krypton 85. Since the radioactive source could only provide at best 5 milliwatts, Brown created an anomalous self running energy device. Brown used an LC oscillator, where the radioactive material ionized a corona around the coil. If the circuit is tuned to the ion-acoustic resonance of the corona, then the ion-oscillations could couple a ZPE coherence directly to the circuit.

Celenza, L.S. and V.K. Mishra, C.M. Shakin, K.F. Liu (1986), "Exotic States in QED," Phys. Rev. Lett. 57(1), page 55; Caldi, D.G. and A. Chodos (1987), "Narrow e^+e^- peaks in heavy-ion collisions and a possible new phase of QED," Phys. Rev. D 36(9), page 2876; Jack Ng, Y. and Y. Kikuchi (1987), "Narrow e^+e^- peaks in heavy-ion collisions as possible evidence of a confining phase of QED," Phys. Rev. D 36(9), page 2880; Celenza,L.S. and C.R. Ji, C.M. Shakin (1987), "Nontopological solitons in strongly coupled QED," Phys. Rev. D 36(7), pages 2144-48.

Cole, D.C. and H.E. Puthoff (1993), "Extracting energy and heat from the vacuum," Phys. Rev. E 48(2), pages 1562-65.

The Super Tube

Cooke, M.B.(1983), Einstein Doesn't Work Here Anymore, Marcus Books, Toronto.

Presents a hyperspace theory where the noble gas nuclei constitute interdimensional node points for channeling energy.

Correa, P.N. and A.N. Correa (1995), "Electromechanical Transduction of Plasma Pulses," U.S. Patent 5,416,391. "Energy Conversion System," U.S. Patent 5,449,989.

Fundamental discovery that a plasma tube tuned to operate at the abnormal glow discharge region exhibits an over unity energy gain. The abnormal glow discharge is a glow plasma that surrounds the cathode just before a vacuum arc discharge (spark) that occurs when the tube is slowly charged with increasing voltage. The abnormal glow exhibits a negative resistance characteristic as the tube begins an arc discharge. The patents illustrate how to make an appropriate charging circuit to cycle the tube in the abnormal glow discharge regime, control the cycle frequency, avoid the losses of the vacuum arc discharge, and rectify the excess energy onto batteries.

Crum, L.A.(1995), "Bubbles Hotter than the Sun," New Scientist 1975, pages 36-40. Summarized in Fusion Facts 6(12), June 1995, page 10.

Overviews recent research in sonoluminescence including Hiller's discovery that the presence of noble gases make the luminosity increase by a factor of 30.

Dufour, J.(1993), "Cold Fusion by Sparking in Hydrogen Isotopes," Fusion Technology 24, pages 205-228.

Everett, H.(1957), "Relative State Formulation of Quantum Mechanics," Rev. Mod.Phys. 29 (3), page 457. Also 1973, B.S. Dewitt N. Graham, The Many Worlds Interpretation of Quantum Mechanics, Princeton University Press.

Everett was the first to introduce a self-consistent formulation of quantum mechanics without invoking postulates regarding the observer. The formulation yields a superspace containing an infinite number of three dimensional universes.

Farnsworth, P.T.(1939), "Cold Cathode Electron Discharge Tube," U.S. Patent 2,184,910.

Cup shaped electrodes provide a cold cathode discharge which has some similarity to the hollow cathode discharge.

Gundersen, M.A. and G. Schaefer (1990), Physics and Applications of Pseudosparks, Plenum Press, NY.

Conference proceedings studying hollow cathode discharge phenomena. If a glow discharge is created in the interior of a hollow cathode, tremendous currents may be switched and triggered by a small external signal. The phenomena generates an intense beam of ions between the electrodes. The initial stage of the discharge exhibits anomalous behavior including a negative resistance characteristic that is a powerful version of the abnormal glow discharge.

Haisch, B. and A. Rueda, H.E. Puthoff (1994), "Inertia as a zero-point field Lorentz force," Phys. Rev. A 49(2), pages 678-694.

Hyde, W.W.(1990), "Electrostatic Energy Field Power Generating System," U.S. Patent No. 4,897,592. The invention is summarized by King (1991).

Kaku, M. (1994), Hyperspace, Anchor Books Doubleday, NY.

Layman's overview to the hyperspace theories of theoretical physics includes general relativity, nuclear standard model, quantum gravity, superstrings, many-worlds, wormholes, time warps, Hawking's universal wave function and Coleman's parallel universes. Shows that unification of physics elegantly occurs via the hyperspace theories. It is noteworthy that the fundamental action in the unifying theories is occurring at the Planck length (10^{-33} cm) setting the stage for a microscopic theory of the vacuum fluctuations.

Kalinin, Yu G. et al.(1970), "Observation of Plasma Noise During Turbulent Heating," Sov. Phys. Dokl. 14(11), page 1074; Iguchi, H.(1978), "Initial State of Turbulent Heating of Plasmas," J. Phys. Soc. Jpn. 45(4), page 1364; Hirose, A.(1974), "Fluctuation Measurements in a Toroidal Turbulent Heating Device," Phys. Can. 29(24), page 14.

King, M.B.(1989), Tapping the Zero-Point Energy, Paraclete Publishing, Provo, UT; ... (1991), "Tapping the Zero-Point Energy as an Energy Source," Proc. 26th IECEC vol.4, pages 364-369; ... (1993), "Fundamentals of a Zero-Point Energy Technology," Proc. Int. Sym. on New Energy, pages 201-217; ...(1994), "Vacuum Energy Vortices," Proc. Int. Sym. on New Energy, pages 257-269.

Lewis, E.H.(1995), "Tornados and Ball Lightning," Extraordinary Science VII (4), pages 33-37; ...(March 1996), "Tornados and Tiny Plasmoid Phenomena," New Energy News 3(9), pages 18-20; ...(Feb 1994), "Some Important Kinds of Plasmoid Traces Produced by Cold Fusion Apparatus," Fusion Facts 6(8), pages 16-17.

Overview of plasmoid phenomena including tornados, ball lightning, and microscopic EV's. Includes an abundant list of references.

Matthey, P.H. (1985), "The Swiss ML Converter - A Masterpiece of Craftsmanship and Electronic Engineering," in H.A. Nieper (ed.), Revolution in Technology, Medicine and Society, MIT Verlag, Odenburg.

McElrath, H.B.(1936), "Electron Tube," U.S. Patent 2,032,545.

Patents the use of radioactive materials to create a cold cathode.

The Super Tube

Michrowski, A.(1993), "Vacuum Energy Developments," Proc. Int. Sym. on New Energy, pages 407-417.

Mizuno, T. and M. Enyo, T. Akimoto, K. Azumi (1993), "Anomalous Heat Evolution from $SrCeO_3$ - Type Proton Conductors during Absorption/Desorption of Deuterium in Alternate Electric Field," 4th Int. Conf. on Cold Fusion. Abstract in Fusion Facts, Dec. 1993, page 30.

Moray, T.H. and J.E. Moray (1978), The Sea of Energy, Cosray Research Institute, Salt Lake City.

Papp, J.(1984), "Inert Gas Fuel, Fuel Preparation Apparatus and System for Extracting Useful Work from the Fuel," U.S. Patent No. 4,428,193; ...(1972), "Method and Means of Converting Atomic Energy into Utilizable Kinetic Energy," U.S. Patent No. 3,670,494.

Pappas, P.T.(1991), "Energy Creation in Electrical Sparks and Discharges: Theory and Direct Experimental Evidence," Proc. 26th IECEC vol. 4, pages 416-423.

Patterson, J.(1996), "System for Electrolysis," U.S. Patent No. 5,494,559.

> The first U.S. "cold fusion" patent with the claim of excess energy (2000%) granted. The Patterson cell contains hundreds of sub millimeter beads made by electroplating multiple, alternating, thin layers of very pure nickel and palladium. It uses a light water electrolyte. It is considered the best of the electrolytic cold fusion type of experiments with complete repeatability and a record gain of over 1000 in one demonstration.

Peat, E.D.(1988), Superstrings and the Search for the Theory of Everything, Contemporary Books, Chicago.

Piantelli, F.(1995), "Energy generation and Generator by means of anharmonic stimulated fusion," International Patent WO 95/20816. Summarized by W. Collis, Fusion Facts 7(6), Dec 1995, pages 14-15.

> Light water cold fusion experiment using pure crystalline nickel that is fully loaded. Once the cell is triggered by a vibrational stress, ultrasonic stationary waves maintain the reaction. The reaction will cease if the temperature rises sufficiently to destroy the crystalline structure of the core.

Prigogine, I. and G. Nicolis (1977), Self Organization in Nonequilibrium Systems, Wiley, NY; Prigogine, I. and I. Stengers (1984), Order Out of Chaos, Bantam Books, NY.

Puthoff, H.E.(1987), "Ground state of hydrogen as a zero-point fluctuation determined state," Phys. Rev. D 35(10), pages 3266-69.

Puthoff, H.E.(1989), "Gravity as a zero-point fluctuation force," Phys. Rev. A 39(5), pages 2333-42.

The Super Tube

Ralphs, J.D.(1993), Exploring the Fourth Dimension, Lewellyn, St. Paul, MN.

Describes how a Euclidian fourth dimension resolves many of the paradoxes in physics as well as begin to explain paranormal phenomena. Uses perception theory from psychology to show how higher dimensions could physically exist without our awareness of them.

Reinhardt, J. and B. Muller, W. Greiner (1980), "Quantum Electrodynamics of Strong Fields in Heavy Ion Collisions," Prog. Part. and Nucl. Phys. 4, page 503.

Resines, J.(1989), "The Complex Secret of Dr. T. Henry Moray," Borderland Sciences, Garberville, CA.

Samokhin, A.(1990), "Vacuum energy - a breakthrough?" Spec. Sci. Tech. 13(4), page 273.

Scheck, F.(1983), Leptons, Hadrons and Nuclei, North Holland Physics Publ., NY, pages 213-223.

Schwinger, J.(1993), "Casimir light: The source," Proc. Natl. Acad. Sci. USA 90, pages 2105-6.

Sego, R.(1981), "The Moray Energy Device - It's Workings," privately published, Provo, UT.

Senitzky, I.R.(1973), "Radiation Reaction and Vacuum Field Effects in Heisenberg-Picture Quantum Electrodynamics," Phys. Rev. Lett. 31(15), page 955.

Shoulders, K.R.(1991), "Energy Conversion Using High Charge Density," U.S. Patent No. 5,018,180.

Fundamental discovery on how to launch a micron size, negatively charged plasmoid called "Electrum Validum" (EV). An EV yields excess energy (over unity gain) whenever it hits the anode or travels down the axis of an hollow coil. The excess energy apparently comes from the ZPE.

Wheeler, J.A.(1962), Geometrodynamics, Academic Press, NY.

By combining the quantum electrodynamic description of the ZPE with general relativity, the author derives the hyperspatial flux (Planck length wormhole) model of the ZPE. The model derives an energy density on the order of 10^{93} grams/cm^3 making it the most powerful of the ZPE descriptions. If true, it could not only offer a limitless energy source, but gravitaional propulsion as well.

Wolf, F.A.(1990), Parallel Universes, Simon & Shuster, NY.

Layman's overview of the various hyperspace theories discussed in today's physics, which suggest an infinite number of three dimensional universes co-exist with our own.

The Super Tube

Charge Clusters:
The Basis of Zero-Point Energy
Inventions

June 1997

Abstract

Many "free energy" inventions utilize (sometimes unwittingly) the phenomena of charge clusters, which may provide the coupling to the zero-point energy (ZPE) for their source of power. Shoulders demonstrates how to produce these micron sized plasmoids, called EV's, and his measurements show they contain a net charge on the order of 100 billion electrons. Recent investigations of the EV suggest its anomalous stability is due to a thin, helical, vortex ring filament possessing an extraordinary poloidal rotational velocity. Such a vortex ring has characteristics similar to the well studied filamentation instability, an intensely tightening, force-free vortex that occurs in non-neutral plasmas; for an EV the filament closes upon itself. The vortex ring can be launched when ions in a highly polarized plasma (or dielectric during breakdown) rush toward a point cathode (or dendrite in a fracturing crystal) whose tip explodes in response to a sharp, electrostatic pulse. EV production is likely occurring in fractoemission, sonoluminescence and in most electrical discharges. If so, it can be the foundation for a unifying hypothesis to explain the energy source behind a wide variety of seemingly diverse inventions including Moray's and Correa's plasma tubes, Grigg's hydrosonic pump, Sweet's conditioned barium ferrite, Brown's electrogravitic capacitors, Gray's motor, Hyde's generator, and even cold fusion. The inventions which utilize fractoemission at their basis tend to retain the EV plasmoids, and their simple construction offers the opportunity for widespread replication.

Introduction

Could there be inventions that tap energy directly from empty space? That depends on the nature of the fabric of space. Today most scientist believe that empty space is simply a void. Across history mankind believed just the opposite. For example ancient Eastern philosophies viewed space as filled with an all-pervading energy called "prana." Since the inception of Western science to the turn of the 20th century, space was thought to be filled with a hypothetical substance called the ether (aether), which would be the medium to support the propagation of light waves. When modeled as a material substance, the ether seemed contradictory. It had to be tenuous to allow matter to pass through it without notice, yet extremely rigid to manifest the high velocity of light. The null result of the Michelson-Morley experiment to detect the ether wind perhaps came as a relief to the scientific community because they could simply accept the postulates of relativity, which yield the principle of Lorentz invariance, without dealing with a cumbersome theoretical artifact that would tend to deny it.

The principle of Lorentz invariance means the state of a physical system (i.e. laws of physics) are the same for all bodies (or observer frames of reference) when they are moving at any constant velocity in free space (an inertial frame). A simplistic, material model of the ether would immediately violate the principle since different observers would experience an ether wind as they move relative to it. Attempts to contrive hydrodynamic models where a material ether is dragged or flows beget complexity as the models are adjusted to recover Lorentz invariance, and in addition the etheric substance must be both tenuous to matter yet rigid for light. The principle of relativity with Lorentz invariance is so fundamental to physics, that the scientific community really can't be blamed for using Occam's razor to excise the belief in a material ether, for a simple void model yields the principle immediately.

Zero-Point Energy

However, the void model cannot explain the nature of light, which was why the ether was invented in the first place. Nor can classical physics explain the spectrum of blackbody radiation, or even why electron atomic orbits don't collapse. The resolution occurred in the 1930's with the theory of quantum mechanics, a mathematical formalism that modeled atomic processes remarkably well. The equations contained a term that represented an underlying, energetic jitter inherent in all processes in nature. The jitter was called the "zero-point fluctuations" (ZPF) or "zero-point energy" (ZPE) since the energy is not thermal and is present at absolute zero degrees Kelvin. Dirac[1] proposed that the source of the jitter was from the fabric of space itself, and it could manifest short-lived pairs of elementary particles, each with their corresponding anti-particle. These were called "virtual" due to their short life times. The discovery of anti-matter and particle pair production in accelerator experiments popularized Dirac's view of space. In the 1940's the theory of quantum electrodynamics[2] was established that heavily relied on the action of the virtual particles to explain atomic interactions with electromagnetic fields. The theory was so successful and accurate, it became the ansatz (pattern) for later theories to model the weak and strong nuclear interactions. In face of such success, do virtual particles really come into physical existence, or are they just an imaginary, theoretical construct?

Rather than have the virtual activity as a theoretical artifact, the theory of stochastic electrodynamics[3] has the zero-point energy at the foundation of all processes in physics, and it shows how quantum behavior arises from it such as blackbody radiation[4] and atomic stability.[5] Recent work has shown how the ZPE may also be at the foundation of macroscopic phenomena such as gravity[6] and inertia.[7] From this perspective, the ether has effectively re-entered physics, not as a material substance, but rather as comprised of highly concentrated fluctuations of electromagnetic field energy.

Stochastic electrodynamics postulates that the ZPE spectral energy density, the function that describes how much energy is stored in space at each frequency, must be Lorentz invariant. There is only one functional form that fulfills this postulate, and that is the energy density must be proportional to the frequency cubed.[3] The function's scaling constant is related to Planck's constant which completely specifies the description of the ZPE spectrum. The spectrum yields accurate results when used in calculations, but there is one philosophical problem: the energy density at every point in space appears to be infinite, since the spectral energy density keeps rising with frequency. Quantum electrodynamics suffers the same problem with its virtual particles, and uses a scheme of "renormalization" to subtract off infinities to leave a finite residue when making calculations. In a practical interaction with matter, the very high ZPE frequencies would be too fast to be absorbed. Nonetheless, the question remains: from where does the zero-point energy come, and how can it appear infinite at every point in space? Likewise, where do these virtual particles come from that pop in and out of space?

Wheeler[8] answers these questions in his "already unified" theory that combines stochastic electrodynamics with general relativity called "geometrodynamics." General relativity shows that when mass or energy density become large, the fabric of space bends into a fourth dimension, and with sufficient density pinches into a hyperspatial form called a "wormhole." The wormhole channels the zero-point energy in the form of electric flux (high density field lines) in a direction orthogonal to our three dimensional space. Flux appears to enter our space through "mini white holes" and leave through "mini black holes." The size of these holes are on the order of the Planck length, 10^{-33} cm. The energy density is enormous, 10^{93} g/cm^3. As the orthogonal flux passes through our space, the mini holes are constantly being created and destroyed in a chaotic maelstrom called the "quantum foam." Vorticity in the flow would manifest as elementary particles. A directional bias in the flow would manifest as vacuum polarization. In this model the elementary particles are like whirlpools sustained by the fourth dimensional flux. Wheeler's geometrodynamics is perhaps the most powerful of the ZPE theories, and if we could find a means to control and coherently channel some of that orthogonal flux into our three space, novel technologies to tap appreciable energy as well as control inertia or gravity might be possible.

Vacuum Polarization

The technical question becomes how can we most influence the ZPE flux? Quantum electrodynamics shows that the different elementary particles have different vacuum polarization interactions with the zero-point energy.[9] In particular, electrons, especially in standard conductors, are effectively in equilibrium with the zero-point fluctuations and manifest the entire interaction as a distributed electron cloud.[10] No net vacuum energy absorption could be expected from standard conductors and antennas. Nuclei on the other hand exhibit steep lines of vacuum polarization converging onto the particle. Whenever nuclei undergo abrupt motion or collisions, appreciable vacuum polarization effects occur yielding precipitation of new particles as well as exotic coherent vacuum states.[11] Plasmas during ion-acoustic resonance have exhibited energetic anomalies[12] whose source could be a coherent macroscopic vacuum polarization[13,14] across the plasma. Recent analysis[15] has shown that the abrupt motion of a dielectric boundary triggers real photons from the ZPE into existence. The energy production is proportional to the fourth derivative of the dielectric's velocity, which is an indication that the motion must be sharply abrupt. Sharp, abrupt motion of matter is a technical key to tapping the zero-point energy.

Sonoluminescence

Eberlein[16] extended her analysis of a dielectric boundary's interaction with the ZPE to explain sonoluminescence, a phenomena where a 50 picosecond light pulse is triggered by a collapsing air bubble in water excited by ultrasonics.[17] An abrupt squeeze and release is provided by the bubble's dielectric boundary which exhibits an appreciable fourth derivative of velocity at the turn-around point (minimum radius) where the surface changes direction. The sharp squeeze and release is similar to the proposal of using sharply pulsed, bucking electromagnetic fields to create a pulsed scalar excitation, which induces a slight "orthorotation" (twisting) of the fourth dimensional ZPE flux into our three dimensional space.[18] Sonoluminescence uses the nuclei in the polarized dielectric, bubble boundary to provide the bucking (scalar) squeeze field. The abrupt, noninertial movement of matter, especially plasma nuclei, appears to be an excellent activator of the vacuum energy.

The trick now is to absorb that energy. Eberlein's study of single bubble sonoluminescence shows that even though the emitted spectrum appears similar to a blackbody radiator at 40,000 degrees Kelvin, it is not a high temperature phenomenon, but rather is an artifact of the ZPE spectrum in its interaction with the bubble boundary. (The artifact might even be expected in view of Boyer's calculation[4] showing how the blackbody spectrum can be derived from the ZPE spectrum). In particular, the lower frequency components of the blackbody spectrum are absent, which result in little interaction with the water to form heat or produce molecular disassociation. The absence of plasma recombination emissions or ion radicals in the single bubble experiments support the thesis that heat is not being produced.[16] Thus, even through we are activating the vacuum energy into light pulses, the light propagates away, unless we arrange for a means to absorb it.

Resonant Absorption

Eberlein's analysis shows that the bubble at its minimum radius (about one micron) does exhibit a resonant interaction with the emissions causing an enhancement by a factor of a thousand. Creating a resonant absorber is another key, especially if the high frequency energy can be converted to a more useful form. Perhaps Mead's[19] nonlinear, microscopic resonators designed to absorb very high frequency ZPE fluctuations and emit low beat frequencies would respond well if it were in contact with the bubbles. Inducing a plasma in the water might likewise provide the opportunity for resonant absorption. Meyer[20] used multiple excitation methods (electromagnetic, light, and ultrasonics) simultaneously to dissociate water, and Puharich[21] electrically stimulated water at the resonant frequency of its hydrogen-oxygen bond to dissociate it. Both inventors induced a plasma in water electrically excited at frequencies in the ultrasonic band and claimed a net energy gain in their systems. Inducing turbulence and cavitation in water might also produce plasma yielding a resonant interaction and absorption of ZPE. Griggs[22] claims to produce heat in excess of his input energy via cavitation in his hydrosonic pump, and Keely[23] claimed excess energy in his water motor based on a turbulent operation similar to water hammer vibrations. Schauberger[24] likewise claimed anomalous energy by causing water to flow in a precessional, turbulent manner through carefully crafted pipes where the flow manifested a characteristic bluish glow akin to sonoluminescence. Also there are cold fusion projects using ultrasonic stimulation producing excess heat.[25,26] Direct evidence for plasma phenomena in water was recorded by Matsumoto[27] where tracks of a small plasmoid (like ball lightning) were photographed. If boundary conditions or turbulence can deform a spherical sonoluminescent bubble toward a pancake shape,[28] the center could pinch and transform it into a plasmoid vortex ring.[29] Plasmoids[30] just might be an ideal resonant absorber of the zero-point energy.

Plasmoids

Plasmoid was the name given by Bostick[31] for a stable, plasma vortex ring he created with abrupt electric discharges. His studies indicated that the plasmoid particles would flow in a helical fashion with the helix closing onto itself producing a toroid. The plasmoid particles undergo a precessional motion (a spin of a spin) with a poloidal spin closing into a toroidal spin. Could such a flow be a natural orthorotator of the four dimensional ZPE flux? DePalma's study[32] of forced precession in a mechanical system exhibited gravitational and inertial anomalies. If the ZPE is the basis of inertia, then it would likewise be the basis of angular momentum. Since systems normally resist precessional motion, a force-free system exhibiting precession would be surprising. If such a system alters the ZPE flux, then it might seem to inertially disconnect from the three space universe, a sort of "Machian" disconnection. (Mach originally proposed the principle that the basis of inertia was the gravitational influence of the "fixed stars"). Moreover such a system would contain excess energy provided from the ZPE flux. Could the ZPE - plasma system have a natural stability in the vortex ring form? Reed[33] has suggested that the mathematical modeling by Beltrami of the force-free vortex might provide the engineering foundation for manipulating the ZPE. The force-free vortex can taper like a tornado and concentrate an extreme energy density toward a point, making it a potentially good activator for ZPE coupling. Such vortex filaments have been observed in plasmas (known as the "filamentation instability"[34,35,36]), and they tend to occur in counter-rotating pairs[37] (Figure 1). Bostick also observed plasmoids to be formed in pairs with opposite helicity arising from plasma turbulence. Could such a model be archetypal and likewise represent vacuum polarization pair production from the turbulent "virtual plasma" of the quantum foam?[38,39,40] The vortex filament from a violent electrical discharge has been observed to sometimes form rings[41] (Figure 2), which has been suggested to be the basis of ball lightning formation. Ball lightning has also been modeled as a vortex ring,[42] and its anomalous energy content suggest that it might be cohering the zero-point energy.[43,44] A careful investigation could occur if we could reliably reproduce ball lightning in the laboratory.

Charge Clusters

It appears that a phenomenon like a miniature form of ball lightning has been discovered and is reproducible in the laboratory. Shoulders[45] has shown that impressing a sharp, negative pulse through a pointed cathode in contact with a dielectric positioned on top of a plate anode (Figure 3) launches a micron sized, negatively charged cluster that he named "Electrum Validum" (EV). Mesyats[46] claimed the same discovery and named it an "ecton." Historically, Le Duc[47] documented this phenomena in his research predating 1900. Shoulders has measured that the one micron diameter EV contains approximately 10^{11} electrons and 10^6 ions. When launched, it typically travels at a velocity about one tenth the speed of light. Shoulders has shown that when an EV strikes an anode it produces an electrical pulse which contains more energy than it took to create it. Moreover, when the EV passes through a coil (or in parallel to a serpentine shaped conductor), a pulse is induced on the coil which also exhibits a greater energy output than the original input. Shoulders and Puthoff[48] suggest that the excess energy is from the ZPE. Mesyats shows that the ecton is launched as a result of an explosive emission from the cathode tip. Just prior to the emission the cathode melts, and an extemely sharp, liquid jet extends from it. The explosion destroys the sharp tip of the electrode, and it is difficult to reuse it to launch another ecton. Shoulders uses liquid metal (e.g. mercury) to maintain the tip and thus keeps it a source for repeatable launching. Shoulders also covers the cathode by a hollow dielectric shroud (Figure 4) designed to separate the EV from all the residual ions in the explosive emission. In addition the shroud might offer another possibility to enhance the effect. Mesyats has shown that a surface plasma on the dielectric amplifies the explosive emission. If Shoulders' shroud could be used like a hollow cathode electrode[49] that accumulates a glow plasma just before firing, the explosive emission through the glow plasma might launch a larger and even more powerful EV.

Experimental support for empowering the EV from a glow plasma may arise from the research of Correa[50] whose tubes are designed to build up a cathode glow which then pulses, a phenomenon known as the "abnormal glow discharge." Correa provides a circuit to restrict the current such that the lossy, vacuum arc discharge that follows is immediately quenched, and causes the tube to pulse at a chosen repetition rate. Correa has measured a significant over unity energy gain from his tubes, and has likely rediscovered the operating principle of the plasma tubes of Moray.[51,52] The research of Mesyats supports the conjecture that the abnormal glow discharge can launch EV's, and these are likely the source of the energy anomalies in inventions that involve electric discharges.

The theoretical stability of EV's presents difficulties. A simple spherical model of 10^{11} electrons would fail due to Coulomb repulsion. Ziolkowski and Tippett[53] proposed a collective plasma system interacting with localized electromagnetic waves. The analysis uses a vacuum polarization displacement current term that manifests a sort of "quantum potential" that compensates for the expansion forces. Such a term is significant only if the formation time of the EV is on the order of a femto second (10^{-15} sec). The analysis involves plasma flow vorticity, but it is not clear what geometry the flow takes. Jin and Fox[54] have proposed a toroidal vortex ring using a classical, nonrelativistic analysis of a nonneutral plasma. They show that a launched spherical charge distribution would not be stable, but rather would naturally pinch into a vortex ring. The analysis attempts to show that the poloidal rotation of the vortex ring could produce a large enough magnetic field to attract the outer circulating electrons in a diocotron[36] slipping stream flow. The poloidal rotational velocity would have to be extraordinary, and the stabilizing ring would have to be an extemely thin filament to keep the poloidal electron velocities well below the speed of light. The calculated energy densities within the filament would be greater than a neutron star. At such energy densities vacuum polarization effects would come into play and should really be included in the model. Also the EV's polarization interaction with the dielectric guide is a stabilizing factor and should likewise be included. A classical, nonrelativistic analysis is probably not sufficient to accurately model the EV. Nonetheless, the suggestion that a thin filament vortex ring is the underlying geometry of the EV can be supported from studying the filamentation instability whose imploding, tightening tendency is recognized to produce extremely high energy densities. Since quantum electrodynamic effects are significant at high energy densities, a complete model of the EV should also include a spectral interaction with the zero-point energy since the thin plasma filament would exhibit a resonant coupling with the short wave length, high frequency, and thus more energetic portion of the ZPE spectrum.

It appears that the ions in the polarized glow plasma and the exploding liquid metal cathode tip combine to create an abrupt compression that forms and energizes the charge cluster vortex ring. Just prior to the explosive emission, the polarized glow plasma and the polarized, liquid metal protuberance form two symmetric, boundary layers surrounding the surface electrons. The explosive emission launches the surface electrons as well as the metal tip, ion boundary layer (Figure 5). The emitted electron cloud then experiences a tremendous explosive compression between the two symmetric ion layers. The metal tip, ion layer punches through the glow plasma layer, creating turbulence which initiates the poloidal rotation of the vortex ring. The perfect boundary symmetry from the liquid protuberance and glow plasma not only causes an abrupt scalar compression which polarizes the vacuum and orthorotates the ZPE flux, but it also provides the needed boundary conditions to create a tight vortex filament that is closed upon itself trapping the excess energy.

Shoulders' observations of the EV's support the conjecture that they are like vacuum energy pumps. As the EV propagates down a dielectric guide, it is constantly ionizing the dielectric surface, emitting electrons and emitting light. Yet the EV does not decay. It yields the same pulse when it hits the anode regardless of the distance it travels in the guide. Also, when the high velocity EV triggers a pulse on a surrounding coil, it likewise does not decay. Moreover, the output pulse from the coil exceeds the input pulse that originally launched the EV. The EV must keep moving to remain stable, and seems to "feed" from its environment, absorbing electrons preceding it and shedding electrons in its wake.

Nuclear Transmutation

When the EV strikes the anode it damages the plate leaving small craters. Some craters exhibit a wider, explosive characteristic and these contain a further surprise: X-ray microanalysis of the crater show a wide range of elements that were not present before the strike implying that the EV induced nuclear transmutations.[55,56] Shoulders has proposed that since the EV can carry about 10^6 ions (or protons in a hydrogen gas experiment), it can accelerate the protons with sufficient kinetic energy to overcome the Coulomb barrier and penetrate the lattice nuclei triggering transmutation events. A classical calculation[57] supports Shoulders' hypothesis. The microanalysis of a palladium target showed magnesium, calcium, silicon, gallium and gold, all from a single EV strike. Shoulders has named the EV carrying nuclei, "Nuclear Electrum Validum" (NEV). The NEV's are created whenever the deuterium loaded palladium (or hydrogen loaded nickel) cracks internally producing the NEV by fractoemission. Fractoemission[58] has been observed to produce light emissions in experiments where a crystal is cracked. The emissions exhibit an anomalous persistence, sometimes for hours. Fractoemission is akin to earthquake lights where a phenomenon like ball lightning arises from an earthquake fissure. The hypothesis that fractoemission of NEV's is the energy source behind cold fusion gains further support from the fact that the anomalous heat in the bulk palladium experiments does not arise immediately after loading the palladium to saturation (it only takes three hours), but rather occurs after hundreds of hours, and then in bursts.[59] Like sonoluminescence, the cracking event is an extremely abrupt microscopic motion. The abrupt motion of the lattice nuclei can produce charged, dendritic edges ideal for EV production. If EV's are produced in the cold fusion experiments, then they might not only be the source behind the transmutation events, but could also produce excess heat directly upon exploding. Thus the source of the excess heat in the cold fusion experiments[60] might be predominantly from the zero-point energy.

Fractoemission

Fractoemission on the interior of a dielectric might offer a means to trap and perhaps directly electrically couple to the EV. Mesyats notes that dielectric breakdown is often a launcher of EV's. In T.T. Brown's electrogravitic experiments[61,62] where a charged capacitor exhibits an acceleration in the direction of the positive plate, the greatest thrusts were observed during dielectric breakdown. Lambertson's[63] ceramic/metallic composite might likewise have a fractoemission phenomena occurring within it to produce excess energy in his experiments. A possible hypothesis to explain the voltage overshoot and excess energy in Hyde's[64] generator might involve fractoemission within the dielectric spacers between the segmented stator elements. Also, Moray[51] used exotic and radioactive materials with his cold cathodes to produce surface, luminescent glow emissions. Could fractoemission also be involved?

The fractoemission plasma hypothesis has been proposed[39] to explain the anomalous energy production from a specially conditioned block of barium ferrite, in Sweet's[65] "Vacuum Triode Amplifier" (VTA), a well-witnessed, seemingly simple, self-running "free energy" device. The hypothesis has gained further support from new information on the process of conditioning the barium ferrite, and offers the opportunity for further replication by others. Sweet's conditioned barium ferrite (typically 6x4x1 inch ceramic blocks) exhibited the peculiar ability to shift its magnetic field in response to a very weak magnetic stimulation. This is unusual because the magnetic domains in barium ferrite do not readily move, especially after being conditioned like a permanent magnet. Another peculiarity was the tendency for the field motion to resonate at 60 Hertz (or any other intended frequency imprinted via the conditioning process). The conditioning steps[39] involved first driving the barium ferrite at 60 Hertz (2 amps) via a coil (600 turns of No. 28 wire) surrounding its perimeter, and then abruptly pulsing the coil from a charged capacitor (6500 microfarads at 450 volts). The pulse is fired at the peak of the 60 Hertz waveform. Also the ceramic block was sometimes sandwiched between charged plates to impress a 20 kilovolt electrostatic potential across it during the conditioning process. No investigator claimed success in replication until one[66] tried a preliminary step: First dip the barium ferrite ceramic block into liquid nitrogen. Then the other conditioning steps could induce multiple cracks in the ceramic, loosening its grains. The easily moved, magnetic fields would occur due to pivoting grain motion (not magnetic domain motion), and such grain motion could be constantly creating and stimulating fractoemission plasmoids trapped within the cracks. (The plasmoids would tend to persist because barium ferrite has a low conductivity and acts like a dielectric.) Sweet used pickup coils above the ceramic to receive the anomalous energy that would be produced when the ceramic block was driven by a weak 60 Hertz magnetic field from the side. If the fractoemission hypothesis applies, Sweet may have discovered an elegant way to magnetically couple to EV plasmoids trapped within a dielectric and extract excess energy from them.

Sparking Devices

Perhaps the most common way to extract energy associated with EV's is to absorb their pulses when they hit the anode. Shoulders has shown that the EV's tend to be the first portion of the plasma launched in discharge events. Correa has shown likewise: the anomalous excess energy only occurs from the anomalous glow discharge, the precursor to the lossy vacuum arc discharge. Pappas[67] has proposed that electric sparks produce excess energy, a hypothesis seemingly confirmed by Shoulders and Correa if the spark is of short duration. Dufour[68] has also observed excess heat produced from sparks in his "cold fusion" experiment involving a pure hydrogen gas tube with stainless steel electrodes. Anomalous energy production seems to also occur in underwater, carbon arc discharge experiments[69] that produce the hydrocarbon fuel COH_2. Newman's[70] large discharging coil and Gray's[71] motor both exhibit abrupt sparking phenomena, and both inventors have claimed excess energy production. The Swiss ML Converter[72] is a famous, self-running energy invention involving counter rotating acrylic disks (segmented like a Wimshurst machine) that produces a bright corona between the disks. Could the corona plasma be comprised of EV's who yield their energy by discharging into the rectifier circuit? Rectifying and efficiently absorbing the energy of large voltage spikes, like those produced by EV anode strikes, is a challenging engineering task. If solved, it could facilitate direct electrical energy conversion in any device where abrupt electric discharges (and EV production) are occurring.

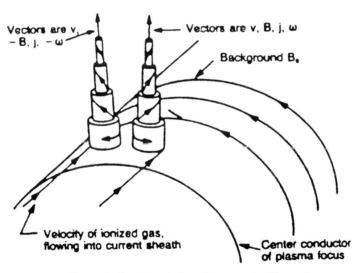

Figure 1 Counter rotating plasma vortex filaments.

A pulse current multiplier (PCM) circuit (Figure 6) has been proposed[52] as a means of converting low power, but extemely high voltage spikes into lower voltage, higher current, wider width pulses that could be rectified by standard circuits. The PCM is similar to a standard voltage divider circuit where a bank of capacitors are charged in series and discharged in parallel. Instead of switches, blocking inductors are used to guide the input voltage spikes down the low impedance, series path. The PCM circuit allows the efficient conversion of weak, electrostatic voltage spikes, which if summed in sufficient quantity, could produce abundant energy as demonstrated by the invention of Hyde.[64] If EV phenomena occur in all electrostatic discharges, then the PCM could provide an efficient method for converting the high voltage discharges and might provide the foundation for a practical device to tap the zero-point energy.

Summary

Despite the beliefs of most of the scientific community, modern physics has accepted that empty space is filled with energetic electromagnetic fluctuations called the zero-point energy. The most powerful model has the ZPE source as a flux from hyperspace. All matter interacts with the ZPE and generally is in equilibrium with it. However, when abrupt matter motion occurs, especially with nuclei, some of the ZPE flux is twisted into our space and can manifest excessive energy. Sonoluminescence may be an example of trapping the energy in a short lived, microscopic resonant structure. The energy may likewise be trapped in microscopic vortex ring plasmoids called EV's. The plasmoid vortex ring might be a natural, resonant orthorotator of the ZPE flux which would cause it to behave like a vacuum energy pump. Since EV's appear to be a precursor within any electric discharge, they may be the source behind the "free energy" machines involving electrical discharges. The excess energy would manifest as high voltage spikes, and an efficient way to convert the spikes to a more useful waveform is by use of a pulse current multiplier circuit. Thus the ZPE coherence induced by the EV could be output directly as electricity.

Long lived EV plasmoids could be produced during fractoemission. They have been proposed to be the cause of excess heat as well as transmutation in the cold fusion experiments. EV's could carry and accelerate ions with sufficient kinetic energy to penetrate the Coulomb barrier of target nuclei, and thus act as microscopic particle accelerators to cause the transmutation of the lattice nuclei. Trapping EV's via fractoemission within the interior of a dielectric might be the most efficient method for extracting their energy since they would tend to persist. Conditioning a barium ferrite ceramic block to maximize internal cracking to promote fractoemission might be the key to replicating Sweet's self-running energy device. If other researchers become successful in reproducing the barium ferrite conditioning, it would set the stage for widespread replication of perhaps the simplest "free energy" device ever invented. Whether EV's trigger pulses on nearby coils, strike anode plates, or get trapped within a dielectric, they appear to provide a unifying hypothesis to explain a wide range of energy devices whose source is apparently from the zero-point energy.

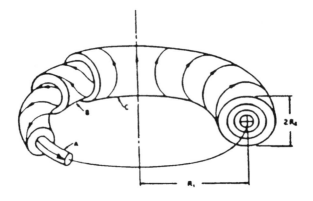

Figure 2. Helical flow in plasmoid vortex ring.

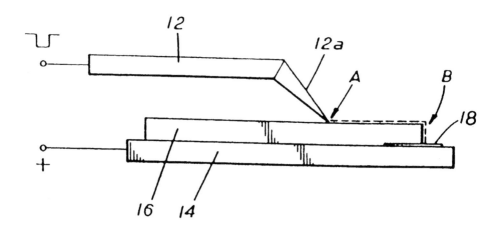

Figure 3 Launching EV plasmoid from sharp pointed cathode (12 a) in contact with dielectric plate (16) on top of metal anode (14).

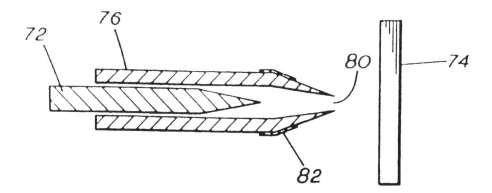

Figure 4 Cross section of cylindrical EV launcher. Dielectric shroud (76) surrounds pointed cathode (72) firing EV toward anode (74). Electrode band (82) is charged to half the anode voltage to repulse ions such that only the negative EV's are launched through the aperture (80).

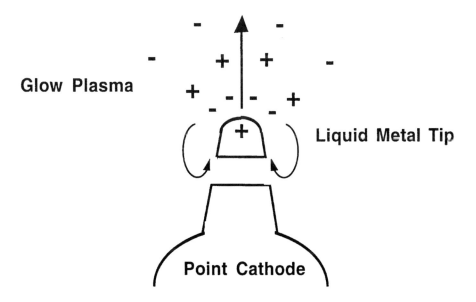

Figure 5 Launching an EV. Ion back-rush helps induce high speed poloidal rotation in the vortex ring.

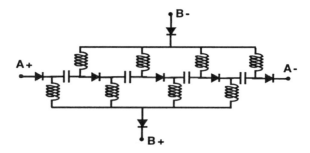

Figure 6 Pulse Current Multiplier (PCM) converts unipolar voltage spikes (input on the A terminals) to current pulses (output on the B terminals). Four stages are shown, but in practice ten or more should be used.

Charge Clusters: The Basis of Zero-Point Energy Inventions

References

1. Dirac, P.A.(1930), Roy. Soc. Proc.126, page 360. Also G. Gamow (1966), Thirty Years that Shook Physics, Double Day, NY.

2. Feynman, R.P. (1985), QED The Strange Theory of Light and Matter, Princeton University Press, Princeton NJ.

3. Boyer, T.H.(1975), "Random Electrodynamics: The theory of classical electrodynamics with classical electromagnetic zero-point radiation," Phys. Rev. D 11(4), pages 790-808.

4. Boyer, T.H.(1969), "Derivation of Blackbody Radiation Spectrum without Quantum Assumptions," Phys. Rev. 182(5), pages 1374-83.

5. Puthoff, H.E.(1987), "Ground state of hydrogen as a zero-point fluctuation determined state," Phys. Rev. D 35(10), pages 3266-69.

6. Puthoff, H.E.(1989), "Gravity as a zero-point fluctuation force," Phys. Rev. A 39(5), pages 2333-42.

7. Haisch, B. and A. Rueda, H.E. Puthoff (1994), "Inertia as a zero-point field Lorentz force," Phys. Rev. A 49(2), pages 678-694.

8. Wheeler, J.A.(1962), Geometrodynamics, Academic Press, NY.

9. Scheck, F.(1983), Leptons, Hadrons and Nuclei, North Holland Physics Publ., NY, pages 213-223.

10. Senitzky, I.R.(1973), "Radiation Reaction and Vacuum Field Effects in Heisenberg-Picture Quantum Electrodynamics," Phys. Rev. Lett. 31(15), page 955.

11. Celenza, L.S. and V.K. Mishra, C.M. Shakin, K.F. Liu (1986), "Exotic States in QED," Phys. Rev. Lett. 57(1), page 55; Caldi, D.G. and A. Chodos (1987), "Narrow e^+e^- peaks in heavy-ion collisions and a possible new phase of QED," Phys. Rev. D 36(9), page 2876; Jack Ng, Y. and Y. Kikuchi (1987), "Narrow e^+e^- peaks in heavy-ion collisions as possible evidence of a confining phase of QED," Phys. Rev. D 36(9), page 2880; Celenza,L.S. and C.R. Ji, C.M. Shakin (1987), "Nontopological solitons in strongly coupled QED," Phys. Rev. D 36(7), pages 2144-48.

12. Sethian, J.D. and D.A. Hammer, C.B. Whaston (1978), "Anomalous Electron-Ion Energy Transfer in a Relativistic-Electron-Beam Heated Plasma," Phys. Rev. Lett. 40(7), page 451; Robertson, S. and A. Fisher, C.W. Roberson (1980), "Electron Beam Heating of a Mirror Confined Plasma," Phys. Fluids 32(2), page 318; Tanaka, M. and Y. Kawai (1979), "Electron Heating by Ion Acoustic Turbulence in Plasmas," J. Phys. Soc. Jpn. 47(1), page 294.

13. King, M.B.(1984), "Macroscopic Vacuum Polarization," Proc. Tesla Centennial Symposium, International Tesla Society, Colorado Springs, pages 99-107. Also (1989), Tapping the Zero-Point Energy, Paraclete Publishing, pages 57-75.

14. Rausher, E.A.(1968), "Electron Interactions and Quantum Plasma Physics," J. Plasma Phys. 2(4), page 517.

15. Barton, G. and C. Eberlein (1993), "On Quantum Radiation from a Moving Body with Finite Refractive Index," Ann. Phys. 227, pages 222-274.

Theoretical analysis of matter in motion interacting with the ZPE. Real photons from the ZPE are created via abrupt motion.

16. Eberlein, C. (1996), "Sonoluminescence as Quantum Vacuum Radiation," Phys. Rev. Lett. 76, pages 3842-45; ...(1996), "Theory of quantum radiation observed as sonoluminescence," Phys. Rev. A 53, pages 2772-87.

Theory that sonoluminescence energy is from a resonant interaction with the ZPE.

17. Barber, B.P. and S.J. Putterman (1991), "Observation of synchronous picosecond sonoluminescence," Nature 353, pages 318-320; ... (1992), "Light Scattering Measurements of the Repetitive Supersonic Implosion of a Sonoluminescing Bubble," Phys. Rev. Lett. 69, pages 3839-42.

18. King, M.B.(1986), "Cohering the Zero-Point Energy," Proc. of the 1986 International Tesla Symposium, Colorado Springs, section 4, pages 13-32. Also (1989), Tapping the Zero-Point Energy, Paraclete Publishing, pages 77-106.

19. Mead, F.B. (1996), "System for Converting Electromagnetic Radiation Energy to Electrical Energy," U.S. Patent 5,590,031; printed in Infinite Energy 2(11), pages 29-34.

System of pairs of microscopic resonators tuned to the high frequency modes of the ZPE. The pairs interact and are tuned to slightly different frequencies so as to emit the beat, difference frequency, which could readily be absorbed by standard circuits. Arrays of microscopic resonators might allow a practical, solid state method to tap the ZPE.

20. Meyer, S.L.(1991), The Birth of a New Technology, Water Fuel Cell, Grove City, OH; ... (1989), "Controlled Process for the Production of Thermal Energy from Gases and Apparatus Useful Therefore," U.S. Patent No. 4,826,581; ... (1990), "Method for the Production of a Fuel Gas, (Electrical Polarization Process)," U.S. Patent No. 4,936,961.

21. Puharich, A.(1981), "Water Decomposition by Means of Alternating Current Electrolysis," Proc. First International Symposium on Nonconventional Energy Technology, Toronto, pages 49-77; ... (1983), "Method and Apparatus for Splitting Water Molecules," U.S. Patent No. 4,394,230.

22. Rothwell, J.(1996), "Notes on the Talk by James Griggs of Hydro Dynamcis, Inc. at the Cold Fusion & New Energy Symposium, January 20, 1996," Infinite Energy 1 (5 & 6), pages 25-27.

23. Pond, D.(1994), "The Keely Motor - How It Works," Proc. Int. Sym. on New Energy, pages 359-371.

24. Alexandersson, O.(1990), Living Water: Viktor Schauberger and the Secrets of Natural Energy, Gateway Books, Bath, UK. Also Frokjaer-Jensen, B.(1981), "The Scandinavian Research Organization and the Implosion Theory (Viktor Schauberger)," Proc. First International Symposium on Nonconventional Energy Technology, Toronto, pages 78-96.

25. Bensen, T.(1995), "A Micro-fusion Reactor: Nuclear reactions in the cold by ultrasonic cavitation," Infinite Energy 1(1), pages 33- 37.

Summarizes the invention of R. Stringham and R. George that uses ultrasonic cavitation to induce cold fusion in palladium foil in heavy water. Excess heat and helium are produced. Electron microscope photos of the foil show pock marks of melted and re-cooled metal.

26. Auluck, S.K.H. and V.K. Shrikande (July 1995), "Proposal for Replication of Stringham and George's Ultrasonic Cavitation Experiment," Fusion Facts 7(1), pages 9-10.

Discusses a pancake collapse model of a sonoluminescent bubble. It can launch a high velocity jet which can inject the contents of the bubble into any solid surface to which it has contact.

27. Matsumoto, T. (1996), "Extraordinary Traces Produced during Pulsed Discharges in Water," Cold Fusion 9, pages 17-21; ...(April 1995), "Artificial Ball-Lightning - Photographs of Cold Fusion," ICCF-5, abstract in Fusion Facts 6(10), pages 24-25; ...(May 1995), "Cold Fusion Experiments by Sparking Discharges in Water," ICCF-5, abstract in Fusion Facts 6(11), page 28.

Observation of tiny ball lightning phenomena in under water discharges for cold fusion experiments. Proposes the formation of hydrogen clusters to induce the fusion events.

28 Prevenslik, T.V.(1996), "Planck Energy and Biological Effects of Ultrasonic Cavitation," Cold Fusion 11, pages 7-10; ...(February 1996), "Biological Effects of Ultrasonic Cavitation," Fusion Facts 7(8), pages 8-9; ...(July 1996), "Ultrasound Induced and Laser Enhanced Cold Fusion Chemistry," Fusion Facts 8(1), pages 15-16.

Model for sonoluminescence has a spherical bubble collapsing into a pancake shape. Zero-point energy modes cohere sufficiently to dissociate water molecules.

29. Bennett, C. (1996), "A Theoretical Mechanism for Sonofusion," Cold Fusion 18, pages 40-42.

Hypothesis that a sonoluminescent bubble collapses into a toroidal vortex ring. Particles accelerated through the center of the vortex have sufficient kinetic energy to generate fusion.

Charge Clusters: The Basis of Zero-Point Energy Inventions

30. Lewis, E.H.(1995), "Tornados and Ball Lightning," Extraordinary Science VII (4), pages 33-37; ...(March 1996), "Tornados and Tiny Plasmoid Phenomena," New Energy News 3(9), pages 18-20; ...(Feb 1994), "Some Important Kinds of Plasmoid Traces Produced by Cold Fusion Apparatus," Fusion Facts 6(8), pages 16-17; ...(October 1996), "The Plasmoid Theory," Cold Fusion 19, pages 37-44; ...(May 1995), "Plasmoid Phenomena," New Energy News 2(12), pages 9-10.

Overview of plasmoid phenomena including tornados, ball lightning, and micro-scopic EV's. Includes an abundant list of references.

31. Bostick, W.H.(1957), "Experimental Study of Plasmoids," Phys. Rev. 106(3), page 404; ...(October 1957), "Plasmoids," Scientific American 197, page 87.

32. DePalma, B.E. and C.E. Edwards (1973), "The Force Machine Experiments," privately published.

33. Reed, D. (1992), "Toward a Structural Model for the Fundamental Electrody-namic Fields of Nature," Extraordinary Science IV(2), pages 22-33; ... (1993), "Evidence for the Screw Electromagnetic Field in Macro and Microscopic Real-ity," Proc. Int. Sym. on New Energy , pages 497-510; ... (1994), "Beltrami Topology as Archetypal Vortex," Proc. Int. Sym. on New Energy, pages 585-608; ...(1996), "The Beltrami Vector Field - The Key to Unlocking the Secrets of Vacuum En-ergy?" Proc. Int. Sym. on New Energy, pages 345-363.

34. Nardi, V. and W.H. Bostick, J. Feugeas, W. Prior (1980), "Internal Structure of Electron-Beam Filaments," Phys. Rev. A 22(5), pages 2211-17.

35. Benford, G.(1972), "Electron Beam Filamentation in Strong Magnetic Fields," Phys. Rev. Lett. 28(19), pages 1242-44; R. Lee, M. Lampe (1973), "Electromagnetic Instabilities, Filamentation and Focusing of Relativistic Electron Beams," Phys. Rev. Lett. 31(23), pages 1390-93; C.A. Kapetankos (1974), "Filamentation of in-tense relativistic electron beams propagating in a dense plasma," App. Phys. Lett. 25(9), pages 484-488.

36. Davidson, R.C.(1990), Physics of Nonneutral Plasmas, Addison-Wesley, NY.

Complete introduction to the topic of nonneutral (or charged) plasmas with abundant references to the scientific literature. The text provides the foundation for the mathematical modeling of nonneutral plasmas.

37. Bostick, W.H.(1966), "Pair Production of Plasma Vortices," Phys. Fluids 9, pages 2078-80.

38. Akimov, A.E. and G.I.Shipov (June 1996), "Torsion Fields and Their Experi-mental Manifestation," Proc. Int. Sci. Conf. on New Ideas in Natrual Science, St. Petersburg Russia.
Summarized by A. Frolov (March 1997), New Energy News 4(11), pages 12-14.

Torsion fields are a model of the physical vacuum similar to Dirac's model except it is comprised of annular wave packets of electrons and positrons instead of electron-positron pairs. Under appropriate conditions the polarized state of the vacuum can be turned into a spin field. Torsion field technology might allow tapping the vacuum energy, inertial propulsion and superluminal communication.

39. King, M.B.(1994), "Vacuum Energy Vortices," Proc. Int. Sym. on New Energy, pages 257-269.

Hypothesis is proposed that counter-rotating plasma vortices can induce a macroscopic pair production of vacuum polarization displacement currents, and such activity yields a large ZPE coherence. An analysis of Sweet's VTA device proposes that fractoemission inside the barium ferrite ceramic (caused by the conditioning process) produces the plasma with grain pivoting inducing the counter-rotation.

40. Winterberg, F.(1990), "Maxwell's Equations and Einstein-Gravity in the Planck Aether Model of a Unified Field Theory," Z. Naturforsch. 45 a, pages 1102-16; ... (1991), "Substratum Interpretation of the Quark-Lepton Symmetries in the Planck Aether Model of a Unified Field Theory," Z. Naturforsch. 46 a, pages 551-559.
A model of the ether comprised of dynamic, toroidal vortex rings.

41. Alexeff, I. and M. Radar (1995), "Possible Precursors of Ball Lightning - Observation of Closed Loops in High-Voltage Discharges," Fusion Tech. 27, pages 271-273.

Closed current loops were photographed during high voltage discharges. The loops enclose a magnetic field of very high energy density. They contract and quickly become compact force-free loops that superficially resemble spheres. In these toroidal geometries, the trapped internal magnetic field balances the external magnetic field to provide an almost force-free configuration. The bibliography cites numerous references on ball lightning.

42. Johnson, P.O.(1965), "Ball Lightning and Self Containing Electromagnetic Fields," Am. J. Phys. 33, page 119.

43. Egely, G.(1986), "Energy Transfer Problems of Ball Lightning," Central Research Institute for Physics, Budapest, Hungary.

44. King, M.B.(1989), Tapping the Zero-Point Energy, Paraclete Publishing, Provo, UT; ... (1991), "Tapping the Zero-Point Energy as an Energy Source," Proc. 26th IECEC vol.4, pages 364-369; ... (1993), "Fundamentals of a Zero-Point Energy Technology," Proc. Int. Sym. on New Energy, pages 201-217.

45. Shoulders, K.R.(1991), "Energy Conversion Using High Charge Density," U.S. Patent No. 5,018,180.

Fundamental discovery on how to launch a micron size, negatively charged plasmoid called "Electrum Validum" (EV). An EV yields excess energy (over unity gain) whenever it hits the anode or travels down the axis of an hollow coil. The excess energy comes from the ZPE.

46. Mesyats, G.A.(1996), "Ecton Processes at the Cathode in a Vacuum Discharge," Proc. 17th International Symposium on Discharges and Electrical Insulation in Vacuum, pages 720-731.

Russian research is presented analyzing their discovery of charge clusters, called "ectons." Ectons often arise from micro explosions on the surface of the cathode, where surface imperfections such as micro protrusions, adsorbed gases, dielectric films and inclusions play an important role. The simplest way to initiate ectons is to cause an explosion of cathode micro protrusions under the action of field emission current. Experiments confirm micro protrusions jets can form from liquid or melting metal. The breakdown of thick dielectric films in their charging with ions also plays an important role in the initiation of ectons. A commonly used way to initiate an ecton is to induce a vacuum discharge over a dielectric in contact with a pointed, metal cathode. An ecton can readily be excited at a contaminated cathode with a low density plasma, but a clean cathode requires a high plasma density.

47. Le Duc, S. (1908), Electric Ions and Their Use in Medicine, Rebman Co., London.

48. Puthoff, H.E.(1990), "The energetic vacuum: implications for energy research," Spec. Sci. Tech. 13(4), pages 247-257.

49. Gundersen, M.A. and G. Schaefer (1990), Physics and Applications of Pseudosparks, Plenum Press, NY.

Conference proceedings studying hollow cathode discharge phenomena. If a glow discharge is created in the interior of a hollow cathode, tremendous currents may be switched and triggered by a small external signal. The phenomena generates an intense beam of ions between the electrodes. The initial stage of the discharge exhibits anomalous behavior including a negative resistance characteristic that is a powerful version of the abnormal glow discharge.

50.Correa, P.N. and A.N. Correa (1995), "Electromechanical Transduction of Plasma Pulses," U.S. Patent 5,416,391; "Energy Conversion System," U.S. Patent 5,449,989; ...(1996), "XS NRG™ Technology," Infinite Energy 2(7), pages 18-38.

Fundamental discovery that a plasma tube tuned to operate at the abnormal glow discharge region exhibits an over unity energy gain. The abnormal glow discharge is a glow plasma that surrounds the cathode just before a vacuum arc discharge (spark) that occurs when the tube is slowly charged with increasing voltage. The abnormal glow exhibits a negative resistance characteristic as the tube begins an arc discharge. The patents illustrate how to make an appropriate charging circuit to cycle the tube in the abnormal glow discharge regime, control the cycle frequency, avoid the losses of the vacuum arc discharge, and rectify the excess energy onto batteries.

51. Moray, T.H. and J.E. Moray (1978), The Sea of Energy, Cosray Research Insti-

tute, Salt Lake City.

52.King, M.B.(1996), "The Super Tube," Proc. Int. Sym. on New Energy, pages
209-221; also Infinite Energy 2(8), pages 23-28.

A powerful plasma tube intended to cohere the ZPE combines the use of a hollow cathode discharge, radioactive cathodes and inert gas mixtures; it operates in the abnormal glow discharge regime. Output energy in the form of large voltage spikes are efficiently absorbed by a pulse current multiplier (PCM) circuit which might offer a solid state means of tapping the vacuum energy.

53. Ziolkowski, R.W. and M.K. Tippett (1991), "Collective effect in an electron plasma system catalyzed by a localized electromagnetic wave," Phys. Rev. A 43(6), pages 3066-72.

Mathematical analysis of Shoulder's EV that includes a significant (vacuum polarization) displacement current term since the EV formation time is of the same order as the plasma frequency period. The resulting nonlinear Klein-Gordon equation contains vorticity terms and a term similar to a quantum mechanical potential which compensates for the repulsion. The system is solved by numerical methods for a stable, localized wave solution which matches the EV in size and charge density.

54. Jin, S.X. and H. Fox (1996), "Characteristics of High-Density Charge Clusters: A Theoretical Model," J. New Energy 1(4), pages 5-20.

A mathematical model of charged clusters (Shoulder's EV's) is presented that shows the stability is due to a helical vortex ring possessing an extraordinary poloidal circulation. In this nonrelativistic calculation, the poloidal filament would have to be thin. A spherical electron cluster is unstable and would tend to form into a toroid by a force balance relationship. The calculation shows that the energy density of the poloidal filament in a charge cluster is a hundred times higher than in a supernova explosion.

55. Shoulders, K. and S. Shoulders (1996), "Observations on the Role of Charge Clusters in Nuclear Cluster Reactions," J. New Energy 1(3), pages 111-121.

Experimental evidence in the form of micrographs and X-ray microanalysis is presented suggesting that nuclear charge clusters, (micron sized plasmoids containing 10^{11} electrons and 10^6 protons or deuterons) can accelerate into lattice nuclei with sufficient kinetic energy to overcome the Coulomb barrier and trigger transmutation events. The hypothesis to explain cold fusion is proposed where electrolytic loading of palladium or nickel causes cracking and fractoemission of the charge clusters.

56. Shoulders, K. and S. Shoulders (1997), "Charge Clusters," Planetary Association for Clean Energy Newsletter 9(1), pages 13-17.

Overview of charge cluster research includes EV's produced by ultrasonics and cavitation charge separation.

57. Fox, H. and R.W. Bass, S.X.Jin (1996), "Plasma-Injected Transmutation," J. New Energy 1(3), pages 222-230.

Classical calculation showing that nuclear charge clusters can be produced at low energy and yet gain sufficient acceleration for their contained protons to penetrate the Coulomb barrier and transmute lattice nuclei.

58. Preparata, G.(1991), "A New Look at Solid-State Fractures, Particle Emission and Cold Nuclear Fusion," Il Nuovo Cimento 104 A(8), page 1259; ... (1990), "Quantum field theory of superradiance," in Cherubini, R., P. Dal Piaz, B. Minetti (editors), Problems of Fundamental Modern Physics, World Scientific, Singapore.

59. Bockris, J.(1996), "The Complex Conditions Needed to Obtain Nuclear Heat from D-Pd Systems," J. New Energy 1(3), pages 210-218.

The hypothesis is proposed that internal cracking of the cathode palladium (or nickel) is the needed triggering mechanism to obtain cold fusion or transmutation events. It explains why, even though it takes only three hours to load palladium rods to saturation, there can be delays of hundreds of hours before heat bursts occur. If the cracks should reach the surface, the deuterium fugacity is diminished and the reaction stops. Thin palladium nickel alloys or layers, as in Patterson's beads, allow the internal cracking to occur quickly giving reliable and repeatable results.

60. Patterson, J.(1996), U.S. Patent 5,494,559. J. Rothwell (1995), "Highlights of the Fifth International Conference on Cold Fusion," Infinite Energy 1(2), pages 8-18.

The first U.S. "cold fusion" patent with the claim of excess energy (2000%) granted. The Patterson cell contains hundreds of sub millimeter beads made in a electroplating process of depositing multiple, alternating, thin layers of very pure nickel and palladium. It runs using a light water electrolyte. It is considered the best of the electrolytic cold fusion type of experiments with complete repeatability and a record gain of over 1000 in one demonstration.

61. Valone, T. (1994), Electrogravitics Systems, Integrity Research Institute, Washington DC.

62. King, M.B. (1976), "Is Artificial Gravity Possible," Tapping the Zero-Point Energy, Paraclete Publishing, Provo, UT, pages 19-32.

63. Lambertson, W.A.(1994), "History and Status of the WIN Process," Proc. Int. Sym. on New Energy, pages 283-288.

64. Hyde, W.W.(1990), "Electrostatic Energy Field Power Generating System," U.S. Patent No. 4,897,592. The invention is summarized in King (1991), reference 44.

65. Sweet, F. and T.E. Bearden (1991), "Utilizing Scalar Electromagnetics to Tap Vacuum Energy," Proc. 26th IECEC vol. 4, pages 370-375; ... (1988), "Nothing is Something: The Theory and Operation of a Phase-Conjugate Vacuum Triode;" ...

(1989), private communication.

66. Watson, D.(1996), private communication.

67. Pappas, P.T.(1991), "Energy Creation in Electrical Sparks and Discharges: Theory and Direct Experimental Evidence," Proc. 26th IECEC vol. 4, pages 416-423.

68. Dufour, J.(1993), "Cold Fusion by Sparking in Hydrogen Isotopes," Fusion Technology 24, pages 205-228.

69.Mallove, E.F. (1996), "AquaFuel and COH_2 Synthesis Gases - More Patents, New Measurements, Speculation..." Infinite Energy 2(10), pages 32-44; "AquaFuel - A Wonderful Fuel, but is it Over-Unity?" Infinite Energy 2(9), pages 44-48.

Carbon arcs underwater producing COH_2 fuel manifests over unity efficiency.

70. Newman, J. (1984), The Energy Machine of Joseph Newman, Joseph Newman Publishing Co., Lucedale, MS. Review by M. Carrell (1996), Infinite Energy 2(7), pages 54-58; special report by E. Soule, pages 58-61.

71. Gray, E.V.(1976), "Pulsed Capacitor Discharge Electric Engine," U.S. Patent 3,890,548.

72. Matthey, P.H. (1985), "The Swiss ML Converter - A Masterpiece of Craftsmanship and Electronic Engineering," in H.A. Nieper (ed.), Revolution in Technology, Medicine and Society, MIT Verlag, Odenburg.

Charge Clusters: The Basis of Zero-Point Energy Inventions

VORTEX FILAMENTS, TORSION FIELDS AND THE ZERO-POINT ENERGY

August 1998

Abstract

An hypothesis is proposed that plasma vortex filaments induce corresponding torsion in the physical vacuum which coheres the zero-point energy. Large energetic effects are expected when the filament is closed into a vortex ring plasmoid (like ball lightning), which can be created by subjecting a glow plasma to an abrupt discharge, bucking electromagnetic (EM) fields, and counter-rotating EM fields.

Introduction

What is the nature of pure vacuum, the "fabric" of empty space? Historically scientists believed space was comprised of a material ether capable of supporting the propagation of light waves. Such a view lead to immediate contradictions: It would have to be stiff as a dense solid to manifest the high velocity of light, yet be tenuous to allow matter to travel through it unperturbed. Failure to detect an ether wind from the earth's relative motion convinced the scientific community to

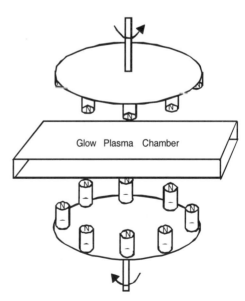

Glow Plasma Chamber

accept the theory of relativity. Here the appealing concept of Lorentz invariance, where all inertial observers (moving at constant velocity) would observe the same laws of physics (including a constant speed of light), yielded a seemingly elegant simplicity: The ether was an unnecessary artifact and could be assumed not to exist. Thus in the early 1900's the scientific community believed empty space to be a void, and today's text books typically support that view.

However, in the 1930's quantum mechanics became accepted because it so accurately described atomic phenomena. The equations of quantum mechanics include a term that describes an ever present, underlying energetic jitter to all phenomena whose source is from the fabric of space itself. The jitter is called "zero-point fluctuations" (ZPF) or "zero-point energy" (ZPE) since the fluctuations are not from thermal radiation, but are present even at absolute zero degrees Kelvin. Quantum electrodynamics (QED) [1] has the ZPE at its basis where the vacuum can manifest pairs of oppositely charged, virtual (short lived) particles. The Heisenburg uncertainty principle is often invoked to allow "borrowing" the ZPE for short time periods in order to explain atomic events and particle interactions. This is a peculiar use of the uncertainty principle since it was originally intended as an inequality expression that limits the accuracy of a quantum measurement, yet in this case it is treated like an equality expression to limit how much energy can be borrowed for how long. Such an approach rationalizes a small "loophole" in the law of energy conservation, which is often cited as the actual reason why the vacuum energy cannot be tapped for large amounts of energy. However, if the ZPE truly physically exists, then real energy is present and its conservation is not the issue.

Zero-Point Energy Paradigm

At the heart of the issue is a paradigm conflict. Most scientists were taught that the vacuum is an empty void, yet quantum theory concludes it's an energetic plenum. It can't be both ways. The conflict has given rise to many paradigm "camps" each with its own characteristic belief regarding the vacuum energy:

Paradigm Camps Regarding ZPE:
1. Quantum physics is wrong. Quantum events can be explained classically using self-fields. ZPE does not exist.

2. Relativity is wrong. A material-like ether exists.

3. Quantum physics is correct, but the ZPE is a theoretical artifact; it is not real.

4. The ZPE physically exists, but its magnitude is too small to be an appreciable energy source.

5. The ZPE physically manifests large energetic fluctuations, but they cannot be tapped because of entropy; they are random and ubiquitous like a uniform heat bath.

6. The ZPE is a manifestation of chaos in an open nonlinear system. Under certain conditions it can exhibit self-organization and therefore become available as a source.

7. The ZPE is a 3-space manifestation of electric flux from a physically real, fourth dimension of space. It can be twisted into our 3-space yielding alterations in the space-time metric. It can be tapped as a source, and doing so locally alters gravity, inertia and the pace of time.

Western academia is dominated by beliefs 3, 4 and 5. Camps 1 and 2 try to model all physical phenomena purely at the classical level. Probably the toughest for them to explain is the EPR [2] nonlocal connection between two separate elementary particles (that originated from a common quantum event). Sophisticated ether models [3] are sometimes proposed, which ironically have similarities to belief 7 whenever the flexible space-time manifold is viewed akin to the ether. The Soviet Union has a large academic interest in beliefs 6 and 7. In addition they have contributed abundantly to the literature involving both theoretical and experimental investigations of torsion fields, a helical spinning coherence in the fabric of space associated with all spinning bodies (including elementary particles). Recently translated articles [4-6] show that many ideas regarding the ZPE considered esoteric and speculative in the West have been taken seriously, theoretically developed, and experimentally investigated by Russian scientists.

There is another major paradigm division across the scientific community regarding hyperspace. Do more physical spatial dimensions exist than the commonly perceived three dimensions of space? (The time dimension does not qualify unless it is considered to be a true spatial dimension as well). No doubt the majority believes not. Physicists often model higher dimensions mathematically, but typically do not believe it represents a real physical hyperspace. Some "compact" the higher dimensions topologically into microscopic tubes so as not to manifest them as spatial dimensions [7]. The philosophical bias is basically "if we can't perceive it, it can't exist." However, the scientific community has always had a minority camp that believes hyperspatial dimensions physically exist [8] and that it is human perception that limits us (like "flat landers") from seeing them. The principle of Occam's razor (simplest theory explaining all) is often cited as the reason hyperspace should not exist. However, if the strange observations of quantum mechanics (e.g., quantum logic, wave particle duality, nonlocal connectivity, etc.) as well as space-time curvature of general relativity are to be physically modeled, then utilizing a physical hyperspace might beget the simplest theory after all. Moreover, such modeling when applied to the ZPE could produce fruitful suggestions for experiments that might tap it for large output power.

Geometrodynamics

Perhaps the most powerful of the ZPE descriptions is Wheeler's geometrodynamics [9] where the (mass equivalent) energy density of a single vacuum fluctuation is on the order of 10^{94} g/cm^3. Wheeler derives this value by inserting the ZPE's spectral energy density expression from quantum mechanics into the stress-energy tensor of general relativity. The high frequency modes of the ZPE yield such a huge energy density that space-time warps (like a black-hole) into microscopic hyperspatial filaments called "wormholes" that can channel electric flux between separate regions of 3-space or possibly to alternate parallel universes embedded in a "super space." The mouths of these wormholes are on the order of the Planck length, 10^{-33} cm (that is twenty orders of magnitude smaller than the electron). In 3-space they manifest like primitive charged particles or "mini holes" whose polarity is set by the direction of electric flux. The electric flux normally passes orthogonally through our 3-space from a fourth dimension with small residual (vectorial) components aligned in 3-space to manifest the ZPE jitter. Thus the flux appears to enter through "mini white holes" and exiting through "mini black holes," and this action creates the foundational basis of charge pair production. The mini hole pairs are constantly being created and annihilated in a fluctuating state of chaotic turbulence called the "quantum foam." On a large scale the quantum foam averages to yield the familiar flat space-time metric, and it provides an ideal substrate to model QED's virtual charge activity. By combining quantum mechanics' ZPE spectrum and general relativity without any extra assumptions, Wheeler created an "already unified" field theory that ironically yields a specific model for the ether when viewed from a 3-space perspective: It appears like a turbulent plasma.

The wormhole model of charged particle pairs offers the prospect of geometric descriptions of the real elementary particles. Such modeling is intuitively appealing especially in view of a similar manifestation that can occur in hydrodynamics, the Falaco soliton [10]. The Falaco soliton is a vortex filament that can be created in a quiescent swimming pool. If a frisbee is stroked across the pool's surface, it will create two oppositely rotating, surface circulations in its wake, which can persist for minutes. An arcing, underwater, thin vortex filament connects the two circulations. If ink drops are poured into one vortex, the ink will flow helically along the arc toward its paired counterpart and highlight the connecting thread vortex filament. If the filament is cut, the Falaco soliton will instantly disappear with an audible pop. In an analogous view, wormhole filaments may provide a way to model nonlocal EPR connectivity where oppositely moving particles born from a quantum pair production event are still connected through the higher spatial dimensions. The vortex filament is an archetype that appears at all levels of nature. It not only plays a role in modeling torsion fields and elementary particles, but it also might well provide the key for abundantly tapping the zero-point energy.

The goal of geometrodynamics is to ultimately model the real elementary particles. Since the ZPE flux flows orthogonally to 3-space, some type of vortex action in this flux, perhaps manifesting as a toroidal vortex ring, is required for the particle's persistence in 3-space. The ZPE flux continuously feeds the vortex in order to maintain it, much as the flow of a river maintains a whirlpool. Geometrodynamics was successful in modeling spin-1 bosons as three dimensional topological entities called "3-geons," but was unable to model the spin 1/2 fermions in a like manner [11]. The problem with spin 1/2 particles is that their isotopic spin description exhibits a quantum two valueness that cannot be modeled geometrically in 3-space. After one 360 degree rotation the fermion does not return to its original state, but instead it alternates between an "up" and "down" state, requiring two full rotations to return to the original state. Thus the electron cannot be modeled as a three dimensional geometric entity. Geometric modeling will require invoking higher dimensions where the electron's space-time manifold becomes "disconnected" from 3-space to allow a fourth dimensional rotation, yet it asymptotically appears to connect to the global space-time metric [12]. From a 3-space perspective the electron would be classed as a "non-orientable" topological entity, like a Klien bottle. (The mouth of a four dimensional Klien bottle appears as a toroid when intersecting 3-space [6,13]. Thus could the Dirac virtual fermion vacuum appear as a sea of toroidal forms? [14]) Hadley [15] has recently proposed a "4-geon" model for the electron using only general relativity, and was able to show that his model yields the strange, non-distributive, quantum logic where the quantum particle seems to somehow nonlocally "sense" its experimental environment to manifest the appropriate conjugate state. The model utilizes a peculiar solution of general relativity call "closed time-like curves" (CTC). The solution involves hyperspatial wormholes (filaments) that can allow a self interaction across the time dimension. This presets the stage for a probabilistic outcome consistent with the environment, even for the "delayed choice" experiments [16]. It is surprising that the strange behavior typical of quantum mechanics can be derived from the classical-like equations of general relativity. The behavior essentially arose from topological vortex filaments spanning across the time dimension.

Plasma Vortex Filaments

Vortex filaments naturally arise in turbulent plasmas. Bostick [17] shows that filaments tend to form in counter rotating pairs. The vortex filament exhibits a force-free, natural flow where the electric (E field) vector and magnetic (B field) vector can both be aligned with the particle velocity vector in an ever-tightening spiral. Moreover, such spiral field behavior can also occur in pure vacuum as well. Kiehn [18] shows that dynamic structures with non transverse E and B fields can occur in vacuum as solutions to Maxwell's equations. They exhibit topological torsion and have spin qualities like the photon. Rodrigues [19] has also derived non-dispersive, solitary wave solutions to Maxwell's equations that can even manifest speeds exceeding light. The Russian literature on torsion fields also describe solutions along the torsion filaments exceeding light speed [4]. (Since light is typically used to "define" the 3-space portion of the metric, such faster than light solutions would seem to propagate on filaments in the hyperspace). There appears to be an archetypal pattern here [20]: Just as turbulent plasmas exhibit a tendency to form into force-free vortex filaments, the virtual plasma comprising the quantum foam can exhibit a like tendency. This suggests a hypothesis: A highly energetic plasma filament induces a corresponding vacuum torsion filament. The coherent tandem of vortex filaments couples to the orthogonal ZPE flux and twists more of it into our 3-space. Since the force-free vortex filament exhibits self-tightening tendency [21,22], a plasma filament which closes onto itself (forming a helical vortex ring plasmoid [23-25]) would experience a positive feed back loop of increasing energy density [26] which ultimately would bring the ZPE effects into play. Once the quantum foam's virtual plasma couples to the vortex ring, the feeding ZPE flux would manifest a "non-orientable" entity akin to a "macroscopic charge." This might be the energy source feeding ball lightning as well as Shoulders' "electrum validum" (EV)[27]. Such vortex phenomena also arises from a plasma's anomalous glow discharge as evidence by the experiments of Correa [28], who appears to have rediscovered the operating principle of the plasma tubes of Moray [29]. Thus the vortex filament manifests in both the virtual plasma of the quantum foam to produce the elementary particles as well as in turbulent plasmas to produce plasmoids, ball lightning and EV's. It appears to be an engineering key for tapping the vacuum energy.

The vortex filament is at the heart of modeling torsion fields where filaments can intertwine and potentially grow as large braids [30]. Shpilman [31] uses this theme to describe the electron as possessing a plethora of vortex filaments flowing from the charge and interacting with its environment. Such a model could likewise be applied to ball lightning or the EV where a dynamic polarization interaction with the dielectric environment [32] seems necessary for its stability. In a description like Shpilman's, an EV sprouts vortex filaments whose negative end is braided around the EV's internal toroid and whose positive end strikes the dielectric substrate, sucking electrons from before it while depositing them in its wake as it travels. Rather than exist as a collection of electrons, the EV could be thought of as a braided cluster of vortex filaments which manifests as a kind of "macroscopic charge." Torsion filaments can be thought of as braids starting with microscopic wormhole filaments and recursively building up in the physical vacuum, even to a macroscopic level.

Models of the Vacuum

The existence of braided torsion filaments in the vacuum requires a rather exotic model for the physical vacuum. The literature contains a variety of proposals for modeling the physical vacuum. Some would support the potential existence of rich set of organized, "subtle" structures, while others would rule them out. In the next century as experimental evidence accumulates, it would not be surprising to see the scientific community divide into paradigm camps regarding how to best model the physical vacuum:

Models of the Physical Vacuum
1. Void
2. Fluid ether
3. Spectral fluctuations
4. Orthogonal electric flux
5. Turbulent plasma
6. Virtual elementary particles
7. Toroids
8. Feynman lattice
9. Phyton lattice

The void model would have to explain action at a distance while a fluid ether would have to explain the observations of physics in a local mechanical way. The zero-point energy density spectrum is popular in the field of stochastic electrodynamics [33] where quantum effects are explained classically by matter's interaction with the all pervading zero-point energy density spectrum. EPR nonlocal connectivity remains unexplained however. Wheeler's geometrodynamics is essentially the orthogonal flux model which manifests a turbulent plasma model in 3-space. Dirac's virtual particle sea [34] was perhaps the first proposal for modern QED with the toroidal vacuum [14] as an attempt to avoid using point particles. The Feynman lattice [35] is a model that tries to explain the manifestations of the wide variety of elementary particles. The phyton lattice features spin and torsion. Since the hypothesis is that vortex filamentation can induce a large ZPE interaction, the phyton lattice model might inspire some effective experimental suggestions.

Akimov [4] is a leading researcher who has proposed the phyton lattice model for the physical vacuum, and this model is popular in

Russia. Akimov hypothesizes that the basis of the vacuum is a quasi-particle called the "phyton" which is sized at the Planck length and exhibits two counter-rotating spins as if it were a pair of particles superimposed one within the other. Obviously for this to be physically modeled, space must contain a higher dimensionality. Thus even though not stated explicitly, Akimov's model is fundamentally hyperspatial in its nature. Phytons proliferate all of space, and for simplicity can be thought of as forming a lattice when in the vicinity of matter. The phyton can polarize in various ways to manifest the fields of nature. It can separate into opposite charge to manifest the electric field and support QED vacuum polarization. The gravitational field is modeled by the phyton separating longitudinally (where the spin axes still remain opposite) in an oscillatory fashion and is called "longitudinal spin polarization." In response to a spinning body one of the phyton spin axis can flip so that both become aligned, resulting in what is called a "spin transverse polarization." Such a polarization is used to model torsion fields and torsion propagation. Akimov's diagrams illustrate how the phyton lattice polarizes in response to a spinning body (including the elementary particles). If the body's spin axis points up, the phytons above the body point up; those below point down. Why that particular polarization? Further discussion is required since that selection relates spin to the higher geometric construct of torsion and chirality. A right hand rule was used to define the spin axis; a left hand rule can have been chosen just as well. By picking this particular spin transverse polarization, Akimov is selecting a preferred chirality for the fabric of space. Chirality refers to spin becoming associated with a helix having either right-handed threads or left-handed threads. Is there a preferred chirality to the fabric of space? Kiehn [36] has described how a chiral vacuum could be topologically modeled in a curved space-time manifold. It would yield subtle electromagnetic and torsional effects. Could this be the basis for preferred molecular chirality observed in living systems? The question is not yet answered, but in Russia there is biological research interest and active experiments exploring the interaction of torsion fields with living systems and possible medical applications [5]. The activation of powerful torsion fields via plasma filaments might produce dramatic effects in the biological arena. Years of Soviet research has garnered experimental support for Akimov's model despite the many still outstanding, theoretical questions. Our understanding of vacuum physics is still at an early enough stage that the science must be

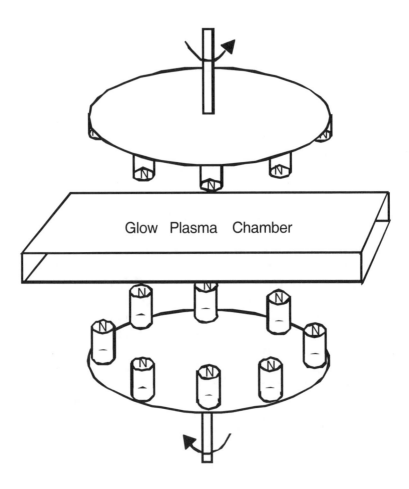

Figure 1 Stimulate a glow plasma with counter rotating, bucking magnetic fields as well as an abrupt electric discharge from a pulsing circuit (not shown).

explored empirically. Perhaps the most valuable contribution would be an experiment that produces such a large energetic effect, that a free running energy machine would be trivial to construct.

Plasma Stimulation

The primary hypothesis for tapping the ZPE is that stimulating a plasma into a self-organized coherent form induces a like coherence in the virtual plasma of the quantum foam. This would suggest that a glow plasma be subjected to one or more of the following stimulations to induce a coherent nonlinear self-organization:

1. Abrupt EM pulse
2. Bucking EM fields
3. Counter-rotating EM fields

There are "free energy" inventions associated with each stimulation. For example Correa [28], Moray [29], Papp [37], Shoulders [27], and Graneau [38] all utilize the abrupt discharge in their inventions to manifest a plasmoid form as well as excess energy [39]. Correa, Moray and Papp apply the discharge directly to a glow plasma. Shoulders pulses a liquid metal electrode and Graneau pulses a small cylinder of water. Shoulders, Correa and Graneau have observed plasmoid formations via photographic methods. Shoulders, Correa and Moray tapped the excess energy via rectifying an output pulse, while Papp and Graneau have focused on tapping the anomalously large mechanical reaction force via a piston. It is noteworthy that an abrupt discharge in a glow plasma or liquid also produces a characteristic bucking field compression. Mesyats [40] has described in detail the behavior of a liquid surface in response to an electric discharge. It forms a stalk protruding from the surface, which is symmetrically surrounded by a polarized glow plasma. The tip of the stalk explodes into the glow plasma yielding a perfectly symmetrical compression by two ion layers. It is the symmetry that guides the emitted electron plasma to form into a closed helical toroidal filament, producing the EV or plasmoid [39].

A number of inventors have used radioactive materials to help create the glow plasma. Moray [29] is perhaps the most famous where his "Swedish stone" cathode contained a mixture of luminescent and radioactive compounds pressed into a germanium pellet to make a "radioactive transistor" with sufficient gain to drive a small loud speaker. Papp [37] also used radium and luminescent materials in his electrodes to help ionize his inert gas mixture. Brown [41] used this principle in a simplified fashion to make his nuclear battery where a weak radioactive source creates a glow plasma which interacts resonantly with an LC circuit to produce anomalously excessive energy output. The simplicity of Brown's invention makes it a good candidate for replication.

The use of bucking EM fields to produce a scalar excitation in the vacuum energy has been emphasized by Bearden [42]. Abruptly opposing EM fields produce a stress on the fabric of space and increase the electric potential, yet since the fields are in opposition, they sum into a net zero field vector. Nonetheless the abrupt stress and release can cause an orthorotation of the ZPE flux [43], and can couple vacuum energy into the glow plasma being so stimulated. Caduceus coils [44] or Mobius coils [45] have been suggested for such excitation. Recent Soviet experiments [6] with Mobius coils have claimed to launch "non-orientable entities" akin to ball lightning, as well as produce negative energy formations described as "magnetic monopoles." Subjecting a glow plasma to abruptly bucking magnetic fields might produce some large ZPE effects.

Bedini [46] used the bucking field principle in his "modified Kromrey, G-field" generator from which experimenters [47] claimed to have observed an output of "cold current," i.e. the ability to conduct appreciable power along thin wires without heating them. Bedini's generator consisted of two steel core coils that are spun through the two air gaps of two aligned horse-shoe magnets. Bedini's plans show ordinary coil windings, but he mentioned using "proprietary windings" as well. Were the windings caduceus, bifilar, or perhaps both? The windings appear balanced and slip rings are used to extract the output which should be combined in parallel such that current would not flow from one coil directly to the other (because of perfect phase opposition). The simplicity of Bedini's invention make it an attractive project for the hobbyist.

Akimov's phyton model motivates using a counter rotation stimulus. If a glow plasma is stimulated with two charged, counter rotating disks, it may activate a significant ZPE coupling. It seems this approach was used for the Swiss ML converter [48,49]. Further excitation could be provided by using the bucking field motif: If magnets are mounted on the surface of each counter rotating disks aligned for repulsion (Figure 1), the glow plasma would be excited by both counter rotation and magnetic field opposition. Gray's motor [50] utilized pulsed opposing magnetic fields along with sparking discharges across the rotor-stator gap. Sweet [51] may have excited a fractoemission plasma by a counter-rotating, pivoting grain motion within his conditioned barium ferrite [52], thus providing stimulation by both techniques. Perhaps an invention that subjects a glow plasma to all three stimulations might produce the largest effects of all.

Summary

Nature appears to operate recursively through levels of self-organization to produce what seems to be an archetypal "quantization" pattern: A plethora of "particles" combine to manifest as a flux or continuous flow at a macroscopic level relative to the size of the individual particle. Nonlinear behavior in the flow can then produce "topological defects" (i.e. solitons, vortices, discontinuities, filaments, vortex rings, etc.), which then can be viewed as if they were "quantized particles" at a new macroscopic level. If similarly gathered as a plethora, they would form the basis for the next higher level flux or continuous flow, which could then repeat the pattern. Turbulent plasmas exhibit the behavior when they form plasmoids. Geometrodynamics proposes analogous behavior to form pairs of charged elementary particles from the quantum foam. Kiehn [53] shows that Cartan's topological methods can abstractly model the archetypal process of evolution of continuous flux to quantized "defects," regardless of the dimensionality. Thus in principle Cartan's methods can apply to hyperspatial models of the ZPE and could someday yield a practical engineering theory for the vacuum energy. When the concept is applied to the substantive models of the physical vacuum, torsion fields become described. Descriptions of wormholes, strings, braids, fiber bundles, vortex rings, etc. appear in the literature, and in a sense could be modeled from a topological foundation of the vortex filament. A primary hypothesis is proposed that triggering a vortical self-organization in a plasma could exhibit a torsional coherence in the zero-point energy especially at the large energy densities that occur as the vortex tightens. The ZPE coupling is maximized when the vortex filament closes into a vortex ring, and the experimental production of excessive energy from ball lightning and EV's may be the evidence. The largest effects might be produced by subjecting a glow plasma to an abrupt discharge combined with a counter-rotating, bucking electromagnetic field stimulation. If such experiments indeed produce large energetic effects, then it would be easy for the scientific community to replicate a self-running device that directly taps the zero-point energy.

Acknowledgments

The author expresses deep appreciation to Neil Boyd and Don Reed for stimulating discussions.

Vortex Filaments, Torsion Fields and the Zero-Point Energy

References

1. R.P. Feynman (1985), QED The Strange Theory of Light and Matter, Princeton Univ. Press, Princeton, NJ; ... (1949), "Space-Time Approach to Quantum Electrodynamics," *Phys. Rev.*, vol 76, p 769.

2. A. Einstein, B. Podolsky, N. Rosen (1935), "Can Quantum Mechanical Description of Physical Reality be Considered Complete?" *Phys. Rev.*, vol 47, p 777.
The original paper which showed that in quantum mechanics particles born from the same quantum event remain strongly correlated even when separated. Experiments in the 1980's confirmed it. The essence of the paradigm shift from classical physics is that the elementary particles cannot be "locally" modeled. See G. Zukav (1979), The Dancing Wu Li Masters, Bantam Books, NY, for a thorough discussion.

3. P.A. La Violette (1985), "An introduction to subquantum kinetics...," *Intl. J. Gen. Sys.*, vol   11, pp 281-345; ... (1991), "Subquantum Kinetics: Exploring the Crack in the First Law," Proc. 26th IECEC, vol 4, pp 352-357.

4. A.E. Akimov (1998), "Heuristic Discussion of the Problem of Finding Long Range Interactions, EGS-Concepts," *J. New Energy*, vol 2, no 3-4, pp 55-80. Also A.E. Akimov, G.I. Shipov (1997), "Torsion Fields and Their Experimental Manifestations," *J. New Energy*, vol 2, no 2, pp 67-84.
Overview of the torsion field research (predominantly in the Soviet Union) includes 177 references. Akimov models the vacuum as a lattice of "phytons," counter-rotating, charged entities sized at the Planck length (10^{-33} cm). Each phyton can polarize in three different ways to manifest 1) electric fields via charge polarization, 2) gravitational fields via oscillating, longitudinal spin polarization, and 3) torsion fields via transverse spin polarization. The gravitational and spin polarizations are hypothesized to support a wave propagation faster than light, and the torsion field supports superluminal links between originally coupled quantum particles to explain EPR "nonlocal" connectivity. Torsion fields can arise from four sources: 1) physical classical spin, 2) the spin of the elementary particles comprising an object, 3) electromagnetic fields, and 4) the geometric form of the object. The spin polarized phyton lattice can also retain a temporary residual torsion field image of a (long standing) stationary object after it is moved.

5. V.F. Panov, et al.(1998), "Torsion Fields and Experiments," *J. New Energy*, vol 2, no 3-4, pp 29-39.
Overviews the history and development of the torsion field research. Summarizes the experimental program at Perm University which includes the effect of torsion fields on chemical and electrochemical processes, crystal growth, nuclear transmutation, biological systems, medical applications, and extraction of energy from the vacuum.

6. Shakhparonov, I.M.(1998), "Kozyrev-Dirac Emanation Methods of Detecting and Interaction with Matter," *J. New Energy*, vol 2, no 3-4, pp 40-45.
Overviews concepts and experiments of creating "non-orientable" topological structures using conductive Moebius band circuit elements. Such topological structures have a hyperspatial nature and cannot be oriented in three dimensional space. The projection into three space manifest phenomena akin to ball lightning. Magnetic

monopoles are likewise of this nature, and experiments are cited where beams of such are launched, detected and shown to exhibit alterations in gravity as well as the pace of time.

7. M. Kaku (1994), <u>Hyperspace</u>, Anchor Books Doubleday, NY.
Layman's overview to the hyperspace theories of theoretical physics includes general relativity, nuclear standard model, quantum gravity, superstrings, many-worlds, wormholes, time warps, Hawking's universal wave function and Coleman's parallel universes. Shows that unification of physics elegantly occurs via the hyperspace theories. It is notewothy that the fundamental action in the unifying theories is occurring at the Planck length (10^{-33} cm) setting the stage for a microscopic theory of the vacuum fluctuations.

8. F.A. Wolf (1990), <u>Parallel Universes</u>, Simon & Shuster, NY.

9. J.A. Wheeler (1962), <u>Geometrodynamics</u>, Academic Press, NY.

10. R.M. Kiehn (1997), "Cartan's Corner," http://www22.pair.com/csdc/car/carhomep;
        ... (1997), "The Falaco Soliton, Cosmic Strings in a Swimming Pool," http://www22.pair.com/csdc/pdf/falaco97.pdf.
Home page of Prof. R.M. Kiehn introduces Cartan's topological methods, which can abstractly model in any number of dimensions a continuous flux or flow. Includes a discussion of the Falaco soliton.

11. C. Misner, K. Thorne, J. Wheeler (1970), <u>Gravitation</u>, W.H. Freeman, NY.
Thorough text on general relativity and differential forms. Chapter 43 discusses the zero-point energy and geometrodynamics. Chapter 44 discusses the problems of modeling actual charge and physics beyond the singularity of gravitational collapse.

12. J.L. Friedman, R.D. Sorkin (1982), "Half-Integral Spin from Quantum Gravity," *Gen. Rel. Grav.*, vol 14, no 7, pp 615-620.
        Describes the topology change required to define the spin 1/2 particles.

13. R.M. Kiehn (1977), "Periods on Manifolds, Quantization and Gauge," *J. Math Phys.*, vol 18, p 4; also http://www22.pair.com/csdc/pd2.
The notions of quantized flux, charge, and spin can be derived from topological ideas of fields built on manifolds. A non Euclidean topology is required for the quantization where the values of the closed integrals on a closed oriented manifold are integer multiples of some smallest value. The flux quantum, charge quantum and angular momentum quantum are not independent but are related by a topological constraint.

14. F. Winterberg (1990), "Maxwell's Equations and Einstein-Gravity in the Planck Aether Model of a Unified Field Theory," <u>Z. Naturforsch. 45 a</u>, pages 1102-16; ... (1991), "Substratum Interpretation of the Quark-Lepton Symmetries in the Planck Aether Model of a Unified Field Theory," <u>Z. Naturforsch. 46 a</u>, pages 551-559.
A model of the ether comprised of dynamic, toroidal vortex rings.

15. M.J. Hadley (1997), "The Logic of Quantum Mechanics Derived From Classical General Relativity," *Found. Phys. Lett.*, vol 10, no 1, pp 43-60; ... (1996), "A Gravitational Explanation of Quantum Mechanics," http://xxx.lanl.gov/abs/quant-ph/9609021.
A pure general relativity solution involving closed time-likes curves is used to model the electron as a four dimensional "geon," which fulfills the non-distributive logic of quantum mechanics. The solution involves a self-interaction across the time dimension (essentially via wormholes) that results in the probabilistic outcomes characteristic of quantum mechanics.

16. J.A. Wheeler (1980), "Beyond the Black Hole," in H. Woolf (ed.), <u>Some Strangeness in the Proportion</u>, Addison-Wesley, Reading, MA, pp 341-375.
Discusses the strange wave-particle descriptions in quantum mechanics. A variation of the two slit diffraction experiments (known as "delayed choice") alter the experimental setup after the particle is launched. Quantum events seem to reach across time.

17. W.H. Bostick (1966), "Pair Production of Plasma Vortices," *Phys. Fluids*, vol 9, pp 2078-80.

18. R.M. Kiehn (1998), "Electromagnetic Wave in the Vacuum with Torsion and Spin," http://www22.pair.com/csdc/pd2.
Torsion and spin wave solutions to Maxwell's equations are derived that are gauge invariant. Values of the spin integral form rational ratios giving a classical solution supporting quantized angular momentum having qualities like the photon.

19. W.A. Rodrigues, J.Y. Lu (1997), "On the Existence of Undistorted Progressive Waves (UPWs) of Arbitrary Speeds (Zero to Infinity) in Nature," *Found. Phys.*, vol 27, no 3, pp 435-508.

20. D. Reed (1992), "Toward a Structural Model for the Fundamental Electrodynamic Fields of Nature," <u>Extraordinary Science IV(2)</u>, pp 22-33; ... (1993), "Evidence for the Screw Electromagnetic Field in Macro and Microscopic Reality," <u>Proc. Int. Sym. on New Energy</u>, pp 497-510; ... (1994), "Beltrami Topology as Archetypal Vortex," <u>Proc. Int. Sym. on New Energy</u>, pp 585-608; ...(1996), "The Beltrami Vector Field - The Key to Unlocking the Secrets of Vacuum Energy?" <u>Proc. Int. Sym. on New Energy</u>, pp 345-363; ... (1997), "The Vortex as Topological Archetype - A Key to New Paradigms in Physics and Energy Science," <u>Proc. Fourth Int. Sym. on New Energy</u>, pp 207-224; ... (1998), "Torsion Field Research," *New Energy News*, vol 6, no 2, p 22; ... (1998), "Excitation and Extraction of Vacuum Energy Via EM - Torsion Field Coupling - Theoretical Model," *J. New Energy*, (to be published).

21. V. Nardi, W.H. Bostick, J. Feugeas, W. Prior (1980), "Internal Structure of Electron-Beam Filaments," *Phys. Rev. A*, vol 22, no 5, pp 2211-17.

22. G. Benford (1972), "Electron Beam Filamentation in Strong Magnetic Fields," *Phys. Rev. Lett.*, vol 28, no 19, pp 1242-44; R. Lee, M. Lampe (1973), "Electromagnetic Instabilities, Filamentation and Focusing of Relativistic Electron Beams,"

Vortex Filaments, Torsion Fields and the Zero-Point Energy

Phys. Rev. Lett., vol 31, no 23, pp 1390-93; C.A. Kapetankos (1974), "Filamentation of intense relativistic electron beams propagating in a dense plasma," *App. Phys.Lett.*, vol 25, no 9, pp 484-488.

23. W.H. Bostick (1957), "Experimental Study of Plasmoids," *Phys. Rev.*, vol 106, no 3, p 404; ... (October 1957), "Plasmoids," *Sci. Amr.*, vol 197, p 87.

24. E.H. Lewis, (1995), "Tornados and Ball Lightning," *Extraordinary Science*, vol VII, no 4, pp 33-37; ... (March 1996), "Tornados and Tiny Plasmoid Phenomena," *New Energy News*, vol 3, no 9, pp 18-20; ... (Feb 1994), "Some Important Kinds of Plasmoid Traces Produced by Cold Fusion Apparatus," *Fusion Facts*, vol 6, no 8, pp 16-17.
Overview of plasmoid phenomena including tornados, ball lightning, and microscopic EV's. Includes an abundant list of references.

25. I. Alexeff, M. Radar (1995), "Possible Precursors of Ball Lightning - Observation of Closed Loops in High-Voltage Discharges," *Fusion Tech.*, vol 27, pp 271-273.
Closed current loops were photographed during high voltage discharges. The loops enclose a magnetic field of very high energy density. They contract and quickly become compact force-free loops that superficially resemble spheres. In these toroidal geometries, the trapped internal magnetic field balances the external magnetic field to provide an almost force-free configuration. The bibliography cites numerous references on ball lightning.

26. S.X. Jin, H. Fox (1996), "Characteristics of High-Density Charge Clusters: A Theoretical Model," *J. New Energy*, vol 1, no 4, pp 5-20.
A mathematical model of charged clusters (Shoulder's EV's) is presented that shows the stability is due to a helical vortex ring possessing an extraordinary poloidal circulation. In this nonrelativistic calculation, the poloidal filament would have to be thin. A spherical electron cluster is unstable and would tend to form into a toroid by a force balance relationship. The calculation shows that the energy density of the poloidal filament in a charge cluster is a hundred times higher than in a supernova explosion.

27. Shoulders, K.R.(1991), "Energy Conversion Using High Charge Density," U.S. Patent No. 5,018,180.
Fundamental discovery on how to launch a micron size, negatively charged plasmoid called "Electrum Validum" (EV). An EV yields excess energy (over unity gain) whenever it hits the anode or travels down the axis of an hollow coil. The excess energy comes from the ZPE.

28. Correa, P.N. and A.N. Correa (1995), "Electromechanical Transduction of Plasma Pulses," U.S. Patent 5,416,391. "Energy Conversion System," U.S. Patent 5,449,989; ... (1996), "XS NRG™ Technology," *Infinite Energy*, vol 2, no 7, pp 18-38; ... (1997), "Metallographic & Excess Energy Density Studies of LGEN™ Cathodes Subject to a PAGD Regime in Vacuum," *Infinite Energy*, vol 3, no 17, pp 73-78.
Fundamental discovery that a plasma tube tuned to operate at the abnormal glow discharge region exhibits an over unity energy gain. The abnormal glow discharge is a glow plasma that surrounds the cathode just before a vacuum arc discharge

(spark) that occurs when the tube is slowly charged with increasing voltage. The abnormal glow exhibits a negative resistance characteristic as the tube begins an arc discharge. The patents illustrate how to make an appropriate charging circuit to cycle the tube in the abnormal glow discharge regime, control the cycle frequency, avoid the losses of the vacuum arc discharge, and rectify the excess energy onto batteries.

29. T.H. Moray, J.E. Moray (1978), <u>The Sea of Energy</u>, Cosray Research Institute, Salt Lake City.

30. J.C. Baez (1992), "Braids and Quantization," http://math.ucr.edu/home/baez/braids.ascii.

31. A. Shpilman, private communications with Neil Boyd.
Shpilman is an Russian inventor who has created technology utilizing torsion fields. His modeling ideas are based on the theme of vortex filaments. Translations of his work might make a significant technological impact. An example: "Spin-Field Generator," http://www.eskimo.com/~billb/freenrg/tors/spin1.html.

32. P. Beckmann (1990), "Electron Clusters," *Galilean Electrodynamics*, vol 1, no 5, pp 55-58.
Explains how Shoulders' EV can be stabilized by a polarization interaction with the nearby dielectric.

33. T.H. Boyer (1975), "Random Electrodynamics: The theory of classical electrodynamics with   classical electromagnetic zero-point radiation," *Phys. Rev. D*, vol 11, no 4, pp 790-808; ... (1969), "Derivation of Blackbody Radiation Spectrum without Quantum Assumptions," *Phys. Rev.*, vol 182, no 5, pp 1374-83.

34. P.A. Dirac (1930), *Roy. Soc. Proc.*, vol 126, p 360. Also G. Gamow (1966), <u>Thirty Years that Shook Physics</u>, Double Day, NY.

35. T. Smith (1997), "The Hyper Diamond Feynman Checkerboard," http://www.innerx.net/personal/tsmith/HDFCmodel.html.
Theory based on Clifford algebra that yields the standard model, masses of the elementary particles, gravity, Bohm's implicate order and the many worlds interpretation of quantum mechanics.

36. R.M. Kiehn (1997), "The Chiral Vacuum," http://www22.pair.com/csdc/pd2.
A chiral constitutive relation is shown to satisfy Maxwell's equations for vacuum and manifests left and right handed helical wave solutions.

37. J. Papp (1984), "Inert Gas Fuel, Fuel Preparation Apparatus and System for Extracting Useful Work from the Fuel," U.S. Patent No. 4,428,193; ...(1972), "Method and Means of Converting Atomic Energy into Utilizable Kinetic Energy," U.S. Patent No. 3,670,494.

38. P. Graneau (1997), "Extracting Intermolecular Bond Energy from Water," <u>Proc. Fourth Int. Sym. on New Energy</u>, pp 65-70; also *Infinite Energy*, vol 3, no 3-4, pp

92-95.
High speed photography reveals a plasma ball sitting in the accelerator muzzle of the water arc explosion experiments.

39. M.B. King (1997), "Charge Clusters: The Basis of Zero-Point Energy Inventions," *J. New Energy*, vol 2, no 2, pp 18-31; also *Infinite Energy*, vol 3, no 3-4, pp 96-102.
Shows how many "free energy" inventions utilize (sometimes unwittingly) the phenomena of charge clusters, which may provide the coupling to the zero-point energy for their source of power.

40. G.A. Mesyats (1996), "Ecton Processes at the Cathode in a Vacuum Discharge," Proc. 17th International Symposium on Discharges and Electrical Insulation in Vacuum, pp 720-731.
Russian research is presented analyzing their discovery of charge clusters, called "ectons." Ectons often arise from micro-explosions on the surface of the cathode, where surface imperfections such as micro-protrusions, adsorbed gases, dielectric films and inclusions play an important role. The simplest way to initiate ectons is to cause an explosion of cathode micro-protrusions under the action of field emission current. Experiments confirm micro-protrusions jets can form from liquid or melting metal. The breakdown of thick dielectric films in their charging with ions also plays an important role in the initiation of ectons. A commonly used way to initiate an ecton is to induce a vacuum discharge over a dielectric in contact with a pointed, metal cathode. An ecton can readily be excited at a contaminated cathode with a low density plasma, but a clean cathode requires a high plasma density.

41. P.M. Brown (1989), "Apparatus for Direct Conversion of Radioactive Decay Energy to Electrical Energy," U.S. Patent No. 4,835,433; ... (1987), "The Moray Device and the Hubbard Coil were Nuclear Batteries," *Magnets*, vol 2, no 3, pp 6-12; ... (1990), "Tesla Technology and Radioisotopic Energy Generation," Proc. Int. Tesla Sym., Colorado Springs, chap 2, pp 85-92.
Brown created a five watt nuclear battery using a weak (one Curie) radioactive source, Krypton 85. Since the radioactive source could only provide at best five milliwatts, Brown created an anomalous self running energy device. Brown used an LC oscillator, where the radioactive material ionized a corona around the coil. If the circuit is tuned to the ion-acoustic resonance of the corona, then the ion-oscillations could couple a ZPE coherence directly to the circuit.

42. T.E. Bearden (1984), "Tesla's Electromagnetics and Its Soviet Weaponization," Proc. Tesla Centennial Sym., International Tesla Society, Colorado Springs, pp 119-138.
Describes "scalar wave" technology. A scalar wave is defined as a propagating organization in the ZPE created by abruptly bucking electromagnetic fields.

43. M.B. King (1986), "Cohering the Zero-Point Energy," Proc. 1986 Int. Tesla Sym., International Tesla Society, Colorado Springs, section 4, pp 13-32. Also ... (1989), Tapping the Zero-Point Energy, Paraclete Publishing, Provo, UT, pp 77-106.

44. W.B. Smith (1964), The New Science, Fern-Graphic Publ., Mississauga, Ontario.

45. S. Seike (1978), <u>The Principles of Ultrarelativity</u>, G-Research Laboratory, Tokyo, Japan.
Explains how a fourth dimensional energy flux can be tapped by pulsed Mobius band coils.

46. J. Bedini (1997), "Collection of Free Energy Machines," http://www.nidlink.com/~john1/index.html.
The modified Kromrey, "G-Field" device was described at a lecture at the 1984 Tesla Centennial Symposium to contain proprietary coil windings which could launch "cold current."

47. T. Bearden, T. Herold, E. Mueller (1985), "Gravity Field Generator Manufactured by John Bedini," Tesla Book Co., Greenville TX.
Experiments showed that Bedini's modified Kromrey, "G-Field" device produced "cold current."

48. P.H. Matthey (1985), "The Swiss ML Converter - A Masterpiece of Craftsmanship and Electronic Engineering," in H.A. Nieper (ed.), <u>Revolution in Technology, Medicine and Society</u>, MIT Verlag, Odenburg.

49. D. Kelly, K. Stevens (1995), "The Swiss ML Converter," *Space Energy Journal*, vol VI, issue 3, pp 13-31.

50. E.V. Gray (1976), "Pulsed Capacitor Discharge Electric Engine," U.S. Patent No. 3,890,548.

51. F. Sweet, T.E. Bearden (1991), "Utilizing Scalar Electromagnetics to Tap Vacuum Energy," <u>Proc. 26th IECEC</u>, vol 4, pp 370-375; ... (1988), "Nothing is Something: The Theory and Operation of a Phase-Conjugate Vacuum Triode;" ... (1989), private communication.

52. M.B. King (1994), "Vacuum Energy Vortices," <u>Proc. Int. Sym. on New Energy</u>, pp 257-269.
Discusses force-free vortices. Suggests that counter-rotating plasmas can induce a ZPE coherence akin to a large scale "pair production."

53. R.M. Kiehn (1997), "Continuous Topological Evolution," http://www22.pair.com/csdc/pd2.
Cartan calculus is used to model continuously changing topology applicable to reversible and irreversible phenomena. It models hole and handle creation/annihilation, turbulence, chaos, action viewed abstractly in terms of kinetic fluctuations, thermodynamics, topological vorticity and parity, harmonic forms and their relationship to angular momentum, probability current, evolution of defects, links and knots, and quantization. The production of torsion defects is the key to the understanding of large scale structures in continuous media. The creation of topological torsion involves discontinuous processes or shocks. The Cartan method is explicitly coordinate free, metric free, and connection free.

Vortex Filaments, Torsion Fields and the Zero-Point Energy

Vortex Filaments, Torsion Fields and the Zero-Point Energy

Transforming the Planet with a Zero-Point Energy Experiment

October 1999

Abstract

A review of some of the best "free energy" inventions of the past 25 years, viewed with the hypothesis that charged clusters couple to the zero-point energy, yields sufficient insight to construct a free-running energy demonstration device. The discussion includes Papp's noble gas engine, Gray's pulsed capacitor discharge motor, Graneau's electric discharge water explosions, Correa's pulsed abnormal glow discharge plasma tubes, Brown's nuclear battery, Hyde's electric generator, Sweet's conditioned barium ferrite, light water "cold fusion" experiments, and

Helical Flow in Plasmoid Vortex Ring Filament

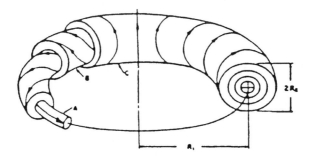

Force Free Vortex Yields Natural Stability

147

Shoulders' charge cluster research. Combining their research reveals a pattern giving the underlying principle for building a self-running device.

Introduction

Can the zero-point energy (ZPE) be tapped as a substantial energy source [1]? The answer depends on the physical nature of the vacuum fluctuations, a topic that begets considerable scientific controversy. The ZPE was first discovered by quantum mechanics as a term in the equations modeling an ever present quantum jitter yielding the uncertainty principle. Quantum electrodynamics (QED) [2] describes it as high frequency, energetic concentrations of electric field energy sufficient to manifest short-lived (virtual) pairs of elementary particles that spontaneously pop in and out of existence. This activity constitutes the substrate of pure empty space. Western academic thought groups into essentially three points of view regarding the paradigm of what constitutes empty space: 1) Space is a void with the ZPE just a mathematical fiction that does not physically exist, 2) only an inconsequential low level ZPE physically exists, or 3) substantial energy is truly present. There is a paradigm conflict between the various schools of thought. The spirit of the Copenhagen interpretation of quantum mechanics represents the first group where the ZPE is to be "renormalized" away without any attempt at interpreting what is underlying the mathematical model. The second group introduces an artificial, high frequency "cut-off" in the equations of the ZPE to limit how much energy can be present, but such a cut-off would then eliminate the energy densities needed for virtual particle pair production, the basis of QED. The third group can argue that the ZPE cannot be tapped appreciably because we have no practical means to interact with the high frequency components, or that the ZPE action is chaotic and it would be an entropy violation to tap it. The last objection can be met via Prigogine's [3] Nobel prize contribution to thermodynamics where he shows how chaotic systems may evolve toward self-organization. Whether a system can be invented to interact with the high frequency ZPE modes [4] puts the debate in the realm of engineering [5]. There is one thing for certain: the only convincing proof that the ZPE can be tapped as an appreciable energy source can only come from an experiment.

There are severe constraints when it comes to an experiment that violates the reigning scientific paradigm. Kuhn [6] shows in his seminal thesis, <u>The Structure of Scientific Revolutions</u>, that paradigm-violating experiments are not welcomed by the scientific community, and often the scientific method regarding their investigation is suspended. Instead political and sociological factors dominate the behavior of the majority of scientists. Typically the experiment is ignored. Rubick [7] has extended Kuhn's thesis to modern times, indicating that if a paradigm-breaking experiment threatens entrenched economic interest as well, there can be a severe and even ruthless suppression of the evidence if the experiment were to begin to successfully replicate. A recent example is the cold fusion class of experiments. Here ridicule is applied, patents blocked, government grants withdrawn, and jobs threatened just over a scientist's willingness to investigate a phenomena that has been replicating world wide as an "underground" activity. If an inventor is lucky (or perhaps unlucky) enough to stumble upon an energetic effect reliable enough to allow the manufacturing of a practical "free energy" machine, suppression tactics have been known to occur such as funding blockage, incessant litigation, patent secrecy orders, personal threats, crime framing, and even assassination. If one wishes to transform the planet with a paradigm-shattering experiment, the key point is to realize that the problem is not a scientific one, but rather a political / economic one. The solution requires completely different tactics than "business as usual."

Kuhn has shown that when the scientific paradigm has changed, it was done by a new generation of scientists willing to study the new phenomena and repeat the anomalous experiments. Thus establishment scientists can be expected to ignore the new phenomena and to ridicule those investigating it. If the experiment begins to replicate, entrenched economic interest can be expected to suppress the discovery. These expectations determine the nature of an experiment that will have a lasting impact.

In particular, the zero-point energy experiment must be:

1) Easy to replicate, so easy that it can be done successfully as a home hobbyist project.

2) Self running. Measurements of excess energy are often difficult

and will be criticized by skeptics no matter what.

3) Easy to describe and share. Information can spread rapidly; a device that has a secret operating principle or requires a special kit can be destroyed and its inventor suppressed.

Inventions with Charge Clusters

There have been a number of successful "free energy" inventions that would have breached the market place if it were not for suppression problems. Some might qualify as candidates for easy replication, but most require considerable effort. There is a pattern to many of the inventions which suggests a common hypothesis regarding how they couple to the vacuum energy: Most have produced (sometimes unwittingly) an abundance of charge clusters [8]. Each cluster is a micron size, charged plasmoid resembling ball lightning that coheres the ZPE to yield excess energy. The hypothesis could concisely explain a variety of "free energy" machines if we adequately understood the nature of the charge cluster itself.

The scientific anomaly essentially regards the charge cluster's existence. Shoulders [9] is credited for the discovery of the charge cluster which he named "electrum validum" (EV). He observed the formation of micron size plasmoids with a net charge measured to be on the order of 10^{11} electrons and 10^6 positive ions. The stability of such a plasmoid is completely anomalous in a classical model because of coulomb repulsion. The best models [10-12] suggest a vortex ring (Figure 1) where an intense toroidal magnetic field causes a tight poloidal spiral (around the toroid) trapping the electrons. Here energy densities similar to a neutron star would be required to compensate the coulomb repulsion. Such energy densities take us out of the classical realm and into quantum electrodynamics where vacuum polarization effects might provide a stabilizing mechanism [13]. The tight toroidal spiral does match behavior often observed in plasma experiments: There is a tendency for plasmas to manifest tight vortex filaments known as the "filamentation instability [14]." Reed [15] shows the vortex matches the force-free Beltrami vortex (Figure 2) where the plasma naturally takes the form of an ever tightening spiral. If such a vortex closes onto itself to make a vortex ring, this tightening tendency propagates around the ring providing positive feedback which squeezes the ring's poloidal radius smaller and smaller until it reaches the dimensions where the high frequency, high energetic components of the ZPE become active and dominant. Thus the natural behavior of the plasmoid itself solves the engineering problem of manifesting boundary conditions on a small enough scale to couple the short wavelength, high energetic modes of the ZPE into the system. The vortex ring model also offers another insight: It might illustrate the mechanism for charge formation directly from the vacuum and suggests that such formation might occur at any scale producing subquantum unstable particles, electrons, micron sized EV's, and ball lightning.

Creating charge clusters requires precise boundary conditions since a linear vortex filament would not naturally close onto itself. Mesyats [16] has shown that "ectons" (his name for EV's) arise from an explosive discharge from a pointed cathode (or from an irregular defect region on the cathode). The instant before the discharge, a microscopic portion of the cathode melts, and a conical stalk of liquid metal protrudes from it (Figure 3). Symmetrically surrounding the stalk is a layer of glow plasma ionization. The EV is formed when the tip of the stalk explodes compressing the surface electrons between two perfectly symmetrical, semi-spherical ion layers (one from the tip, the other from the glow plasma). It is the perfect conical symmetry of the liquid stalk that provides the precise boundary conditions needed to form the vortex ring as a single complete entity. Thus EV formation can be enhanced by using liquid metal tip electrodes like Shoulders suggests (Figure 4) or

Beltrami Vortex
Key To Unlocking Vacuum Energy

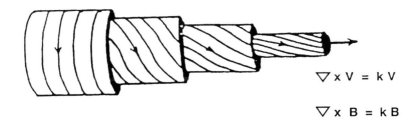

$$\nabla \times V = k V$$

$$\nabla \times B = k B$$

Tightening, Force Free, Vortex Filament

by first inducing abundant cathode glow plasma as Mesyats describes. Forming the liquid or plasma stalk is the key phenomenon that underlies EV creation.

Shoulders suggests two methods for tapping the excess energy from an EV. Either rectify the electric pulse that is produced when the EV hits the anode, or use a pickup coil surrounding a traveling wave tube through which the EV accelerates. Most inventions that have manifested excess energy from plasma EV activity have used the rectification technique. For his 50 KW device Moray [17] invented special rectifier tubes he called "valves" to store the energetic pulsations from his ion oscillator tubes onto capacitors, which then become the input to the next stage of his device. Correa [18] rediscovered Moray's fundamental operating principle in his pulsed anomalous glow discharge (PAGD) tubes. Here, rectification is used to recharge batteries for over unity power gain. Correa shows how to launch just the spark precursor (which contains the excess energy) and quench the lossy arc that follows by stopping the current flow. Correa limits the current by a resistive feeder circuit whereas Moray used capacitor discharge to control the feeding current pulses. Correa precharges the plates with glow plasma from a voltage source whereas Moray used a radioactive cathode to build up glow plasma to trigger it at a lesser voltage. Brown [19] likewise used a radioactive cathode in his resonant nuclear battery claiming one of the simplest embodiments of Moray's original discovery. Correa's and Brown's inventions are excellent candidates for replication at a university facility.

If every spark, even weak electrostatic discharges, contained an EV precursor, then Wimshurst like devices that produced a plethora of electric discharge pulses, where each is rectified and capacitively stored, could sum them to produce a surprising energy gain. The Swiss ML converter [20,21] might be based on this principle as well as Hyde's [22] electrostatic field chopper. Hyde might have produced EV discharges across the dielectric separators between adjacent stator segments to produce the ten-fold voltage spike he was observing. Hyde meticulously constructed hundreds of small, voltage step down, rectifier circuits connecting each stator segment to its distant counterpart across the device, and then further summed their output to produce a net 20 KW DC output while self running. That is an extraordinary claim for such a simple device (in principle). If Hyde's claims are true, then mechanical means can be used to induce an abundance of small EV discharges (which are rectified and summed) to yield a powerful and robust energy machine.

Abrupt electric discharges have been observed to produce an anomalous mechanical reaction force. Graneau's [23,24] experiments with capacitor discharge, water arc explosions have demonstrated both excessive force and over unity energy production. He also observed with high speed photography the formation of a ball lightning plasmoid in the chamber associated with the explosive event. It was important to discharge the capacitor rapidly to induce the anomalous event. His experiments showed that with the same capacitor energy if the discharge pulse was not sharp enough, no explosion occurred, and the water simply remained in the chamber. Anomalies associated with the abrupt transition from matter to plasma have also been observed in recent experiments [25] of firing a femtosecond pulsed laser into xenon to produce plasma clusters (or perhaps EV plasmoids) that resulted in anomalously excessive kinetic energy of the xenon ions. Perhaps the most famous energy invention to take advantage of this phenomena is Papp's [26] noble gas engine. Papp filled a mixture of the inert gases into a sealed piston that gets fired with an electric arc (aided by radioactive electrodes). The resulting explosion drove the piston and produced anomalously excessive energy. Papp claimed the inert gas was "the fuel" which never gets consumed. Today the hypothesis is that any abrupt transition from matter to plasma coherently activates the ZPE, and that such transitions are likely to create EV plasmoids. This hypothesis could be the foundation for creating many future inventions that tap the zero-point energy.

A noteworthy invention that may have utilized both the electrical and mechanical energetic aspects of EV's is Gray's [27] pulsed discharge electric engine. Gray energized opposing electric magnets by firing a capacitor discharge across the air gap between pairs of rotor coil magnets and stator coil magnets timed such that pulsed magnetic repulsion would drive the rotor. The circuitry was designed such that the pulse from the capacitor would energize the magnets and then return to a rectification circuit that would recharge the battery. At first analysis, engineers thought that the arcing in the air gap would foolishly produce tremendous losses, but Gray demonstrated with his prototypes that not only did the motor produce a large torque, it ran so efficiently that the windings and motor housing remained cool. For his invention Gray won the inventor of the year award in 1976, but he ran afoul of the Securities and Exchange Commission for his claims of over unity efficiency. If every electric discharge across the rotor gap abundantly produced charge clusters, the motor may have actually tapped the ZPE as its energy source. Today, a replication of Gray's invention (as well as Correa's) could use large (e.g. one farad) commercial capacitors in place of the batteries to make the device self running.

If charge clusters are a conduit to the vacuum energy, it might be more efficient to trap and couple to them rather than rectify their decay pulse. The phenomena of fractoemission [28] might offer a method where EV's could be created and their energy extracted all within a solid state device. Fractoemission occurs when a crystalline material cracks and forms an anomalously persistent glow plasma in the crack. In some experiments the luminescence has been observed to persist for hours. Fractoemission is likely a small scale occurrence of the same phenomena as earthquake lights, where ball lightning plasmoids are emitted from an earthquake fissure. If fractoemission produces EV's, it could explain both the excess energy and transmutation anomalies associated with the cold fusion experiments. In the cold fusion experiments palladium or nickel is electrochemically loaded with hydrogen or deuterium to the point of saturation via electrolysis of water containing lithium hydroxide (or various salts). Best results are obtained by using a pure crystalline hydride. For a skilled electrochemist it only takes a few hours to completely load the hydride, but the heat anomaly typically manifests only after days of DC and pulsed electrical excitation [29]. After a time, the hydride begins to form cracks and then the anomalies are observed. If the cracking produces fractoemission EV's, and these couple to the ZPE, then both the excess heat and the lattice element transmutation can be explained.

Shoulders [30] demonstrates the EV transmutation phenomena by a trivial experiment: Just fire a single spark from a small Tesla coil onto an pure aluminum plate. With x-ray micrographic analysis determine what elements are formed in the crater where the spark struck. In his studies Shoulders has shown that EV strikes produce transmutation of the anode metal. The predominantly negatively charged EV accelerates toward a nucleus in the anode lattice. It also carries with it some residual ions which would smash into the nucleus like ions launched from a particle accelerator. Moreover, if the residual ions circulate around the EV vortex ring, the ring's symmetry would tend to aim its center (hole) straight at the nucleus, and the ions circulating through the center would be aimed directly at the nucleus with further force. Many cold fusion researchers have observed the transmutation anomaly. Patterson's [31] beads consisting of electroplated layers of nickel and palladium could perhaps be the most reliable and repeatable of the cold fusion, light water experiments. Here fractoemission seems to readily and quickly occur yielding both the heat anomaly as well as transmutation. Unfortunately the cracking could also tend to make the beads flake and eventually wear out. Nonetheless Patterson's beads make available the means for anyone to successfully replicate the transmutation and excess heat anomalies in a light water, cold fusion type experiment.

It would be ideal if a means could be discovered to induce fractoemission EV's, trap them within the material rather than dissipate them as heat, and then tap their energy directly via electromagnetic coupling. The invention of the late Floyd Sweet [32] might be just that discovery. Sweet specially conditioned two, 6 x 4 x 1/2 inch ceramic blocks of barium ferrite such that the magnetic field of each would easily vibrate in response to a weak AC signal from nearby, air core, electromagnetic "exciter" coils (Figure 5). This behavior itself appears anomalous since barium ferrite is a permanent magnetic material, and its magnetic domains do not easily shift. A bifilar pickup coil (the wire consists of two, twisted, insulated strands) sandwiched between the ceramic blocks would pickup an anomalously large output signal in response to the stimulated magnetic vibrations. Moreover, the induced current flow in the bifilar pickup was in opposite directions on the two strands. Perhaps this is the strangest of the observations since it rules out standard magnetic induction. In addition, the output current was observed to be "cold" where many amps could be conducted into a load through thin wires without heating them. Some of the output signal was fed back to the exciter coils to make the device self running. This feedback had to be done carefully for if the ceramic blocks were over driven, they would crumble into powder or dangerously explode. Sweet's invention is spectacular because from such apparent simplicity arises incredible behavior including excessive energy.

How can the invention be explained? Sweet was secretive regarding the conditioning process, but recent information from other researchers

[33] combined with the fractoemission hypothesis might allow its rediscovery. The first clue is the easy magnetic field shifting exhibited by the ceramic blocks. Since barium ferrite's magnetic domains do not readily move, it was proposed [34] that microscopic grains within the ceramic were cracked free by the conditioning process to allow them to easily pivot, and that such motion would produce and stimulate fractoemission EV plasmoids. If the plasmoids are produced around the peripheral region of the block, and if each energetically couples with its neighbor, then it is possible to induce an energetic vacuum polarization (or perhaps torsion field [35-37]) vortex in each block such that the top block contains a circulation opposite to that in the bottom block. Alternatively each single block could support a dual circulation since Sweet did have early successful embodiments that used only one ceramic block. Counter rotation is a principle for inducing any type of coherent pair production from the vacuum energy whether it be virtual elementary particles or macroscopic torsion fields. QED shows the principle arises from conservation of angular momentum. It is the counter rotating vortices that induce the energy onto the bifilar pickup coil bidirectionally, and its strands are guiding vacuum polarization currents instead of standard electron conduction. The last point might explain why the output wires remain cold. Sweet's invention exhibits so much new phenomena associated with the zero-point energy that its rediscovery would accelerate science into a new paradigm. Furthermore, its apparent simplicity makes it the ideal candidate for a repeating experiment that can be done even at home.

Conditioning Barium Ferrite

The goal is to create many charge clusters via fractoemission while trapping them within the material, and then to extract any excess power they might produce via a bifilar pickup coil. If we can induce a resonant, energetic circulation inside the material from an orderly alignment and mutual coupling of charge clusters located around the ceramic's periphery, a pair of macroscopic ZPE vortices might be created. The conditioning process is designed to make freely moving grains within the ceramic. As the grains rock back and forth in phase with a weak A.C. magnetic field from the nearby exciter coils, they create and stimulate fractoemission plasmoids in the microscopic cracks between the movable grains and the stationary portion of the ceramic body. Since Sweet attempted many variations of coil sizes and positioning in his experiments (most of which produced some degree of success), the exact parameters are unclear. The coil descriptions reflect the range as described by researchers close to the project. The conditioning steps are as follow:

1. Freeze the material. This will cause it to readily crack. Ideally a home freezer would be sufficient. If not, pack the ceramic block in dry ice. If it still fails to crack, dip it in liquid nitrogen, but be careful: It should be cold enough to create microscopic cracks, but not so cold as to shatter the entire ceramic.

2. Sandwich the block between two metal plates charged as a parallel plate capacitor at a high DC potential (20 KV) to induce interior corona (Figure 6). It should remain here during the steps that follow.

3. Stimulate the block with 60 Hz bucking magnetic fields from a bifilar coil around its perimeter (240 to 400 windings, 20 gauge, 1 to 2 amps). The coil can be the same one used for receiving the output when the device is running. Sweet claimed that the frequency of this conditioning signal would determine the natural resonant frequency of the self running device. It will be interesting to investigate why this "scalar" excitation is necessary. (In some embodiments two smaller diameter, bifilar coils were placed on the face of the ceramic block instead of perimeter windings).

4. Abruptly pulse the block with a huge magnetic pulse from a standard wound, perimeter coil (220 windings, 14 gauge) by discharge from a large capacitor (6500 microfarads, 450 volts). The ideal timing would be at the peak of the 60 Hz signal driving the bifilar coil. (Be careful here, an electric shock could be lethal). If the barium ferrite block is already a permanent magnet (with one face north, the other south), then the magnetic pulse should be of opposite polarity. The pulse will flip the domains at the ceramic's periphery nearest the coil windings. If the abrupt pulse fails to loosen any grains, turn the block over and repeat the conditioning.

The last step is intended to create many microscopic cracks near the ceramic's periphery to yield freely moving grains. Clearly some trial and error is required to get the conditioning right. If the ceramic cracks too much, it could crumble, or the cracks would be too wide to support fractoemission. If not cracked enough, there could be no grain motion at all. The following suggestions will help the investigation of proper conditioning:

1. Work with small symmetric pieces at first. Many blocks could easily crumble or crack too much during the trials to find the proper stimulation that makes only microscopic cracks.

2. If grains are moving easily, the moving magnetic field can be readily observed with magnetic "green paper," a commercial plastic film designed to image magnetic fields. The image should easily move like a large magnetic bubble in response to a weak stimulating magnetic field from a small, hand held magnet brought nearby. An unconditioned ceramic would not manifest such motion since the domains of a permanent magnet do not easily shift. This comparison would be an indication of success; an unconditioned ceramic can act as a control sample for the experiment.

3. If a permanent barium ferrite magnet is used where one face is north and the other south, the magnetic field on the surface should be as uniform as possible to get the best results. Before conditioning a block Sweet would carefully measure and map the magnetic field on its surface hoping for a deviation of less than five percent. A uniform block will condition symmetrically and thus is more likely able to support the peripheral energy circulation.

4. A surprising energetic activity should be observable on an oscilloscope from a pickup coil placed directly above the material. Fractoemission plasma is known to persist for hours. A freshly conditioned ceramic could manifest electrical activity even in the absence of grain motion. When grain motion occurs an anomalously large pickup would then be observed. This would constitute confirmation of the fractoemission hypothesis.

If the fractoemission hypothesis is correct, it opens the possibility for a host of other experiments using other materials. Barium ferrite was selected because it is a dielectric, and can thus trap the fractoemission plasma. Typically magnetic materials are conductive and thus they would dissipate any charge clusters produced. However, other permanent magnetic materials might work as long as they are electrical insulators and could be successfully conditioned to have moving grains. Generalizing further, any dielectric material (not necessarily magnetic) could be used as long as the grain motion that stimulates the fractoemission plasma can be controlled (e.g. an electret or piezoelectric material where an external electric field controls the grain motion, or even a neutral ceramic where sonic means are used to synchronize grain motion). Of course in each case the trick would be to properly crack the material into having freely movable grains that induce fractoemission. Confirming the fractoemission hypothesis sets the stage for discovery of a wide range of solid state devices that could tap the zero-point energy.

Summary

The empirical studies of Shoulders and Correa have brought zero-point energy research to a new plateau where the hypothesis of charge cluster EV formation could be the key for successfully creating zero-point energy machines. By examining many energy inventions with this hypothesis in mind, a pattern arises for techniques to extract the EV energy. Most inventions rectify the EV pulses as they decay. Some mechanically react to an explosive event from abrupt transitions from matter to plasma that likely form EV plasmoids. Some extract the heat the EV's produce when they decay. Possibly, the most efficient means might be to trap the EV's in a material and couple to them. Fractoemission might be a means to produce and trap the EV plasmoids, and the invention of Sweet might qualify as the most elegant embodiment of a free energy device in the history of the field.

Sweet's discovery has the potential to become a paradigm shifting experiment because its apparent simplicity could allow wide spread replication. It all depends on rediscovering the barium ferrite conditioning process. If it becomes easy to create freely moving grains in even small pieces of the ceramic, and if such grain motion does produce some energetic activity, then it is only a matter of engineering to increase the effect to where an appreciable self running demonstration can be invented again. Once the fractoemission, EV hypothesis is experimentally confirmed, other means can be engineered to create reliable and robust energy machines.

Perhaps most importantly, the discovery of a paradigm shifting experiment will require a team effort, for the experiment must replicate widely to have a lasting impact, and this will require willing cooperation among researchers to help each other succeed. It is our harmonious participation that will manifest a zero-point energy experiment capable of transforming the planet.

Acknowledgments

The author wishes to thank Walt Rosenthal and Don Watson for informative discussions.

Helical Flow in Plasmoid Vortex Ring Filament

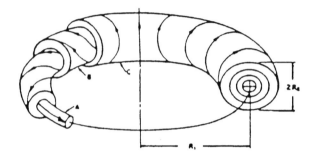

Force Free Vortex Yields Natural Stability

Figure 1 Charge cluster modeled as a light vortex ring plasmoid. The thin ring coherently interacts with the zero-point energy and might be the archetype for charge formation from the vacuum at a variety of sizes ranging from elementary particles to ball lightning.

Beltrami Vortex
Key To Unlocking Vacuum Energy

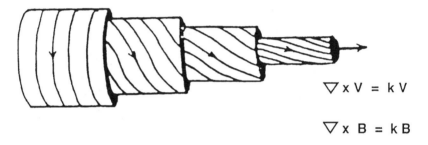

$$\nabla \times V = k\,V$$

$$\nabla \times B = k\,B$$

Tightening, Force Free, Vortex Filament

Figure 2 A plasma filament exhibits a natural force-free vortex which tightens because the flow aligns with its own vorticity (curl) vector. The tightening vortex yields an extraordinary energy density at the tip which activates the zero-point energy.

Explosive Emission

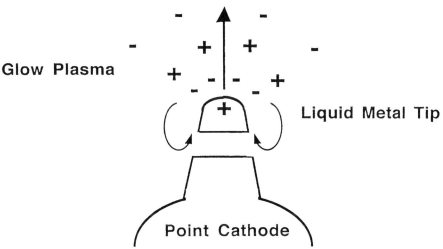

Figure 3 A vortex ring plasmoid is created from an explosive emission from a liquid metal stalk protruding from the cathode. The tip of the stalk explodes into the surrounding glow plasma. The perfect symmetry of this compression event yields a closed vortex filament.

Pure EV Launcher

(Cross Section of Cylinder)

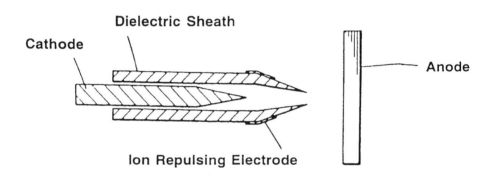

Figure 4 Shoulders can repeatedly launch charge clusters from a liquid metal cathode. The dielectric sheath guides a layer of conductive liquid to the cathode point. An ion repulsing electrode helps to prevent residual ions from contaminating his pure "electrum validum" (EV).

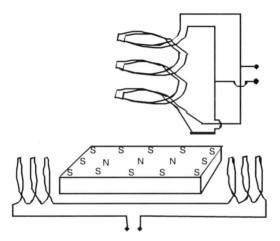

Figure 5 Sweet's conditioned barium ferrite has peripheral domains aligned opposite to those inside. Excitor coils on side vibrate grains to induce oppositely flowing currents on bifilar pickup coil from internal fractoemission plasma activity.

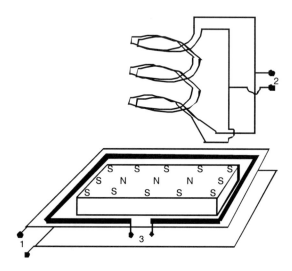

Figure 6 Conditioning barium ferrite with three excitations: 1) DC on the parallel plates above and below the ceramic block, 2) scalar AC on the bifilar coil with currents in opposition, and 3) huge pulse on the perimeter coil from a capacitor discharge.

Transforming the Planet with a Zero-Point Energy Experiment

References

1. D.C. Cole, H.E. Puthoff (1993), "Extracting energy and heat from the vacuum," *Phys. Rev. E*, vol 48, no 2, pp 1562-65.

2. R.P. Feynman (1985), <u>QED The Strange Theory of Light and Matter</u>, Princeton Univ. Press, Princeton, NJ; ... (1949), "Space-Time Approach to Quantum Electrodynamics," *Phys. Rev.*, vol 76, p 769.

3. I. Prigogine, G. Nicolis (1977), <u>Self Organization in Nonequilibrium Systems</u>, Wiley, NY;
   I. Prigogine, I. Stengers (1984), <u>Order Out of Chaos</u>, Bantam Books, NY.

4. F.B. Mead (1996), "System for Converting Electromagnetic Radiation Energy to Electrical Energy," U.S. Patent 5,590,031; printed in *Infinite Energy*, vol 2, no 11, pp 29-34.
   System of pairs of microscopic resonators tuned to the high frequency modes of the ZPE. The pairs interact and are tuned to slightly different frequencies so as to emit the beat, difference frequency, which could readily be absorbed by standard circuits. Arrays of microscopic resonators might allow a practical, solid state method to tap the ZPE.

5. M.B. King (1989), <u>Tapping the Zero-Point Energy</u>, Paraclete Publishing, Provo, UT; ... (1991), "Tapping the Zero-Point Energy as an Energy Source," <u>Proc. 26th IECEC</u>, vol 4, pp 364-369; ... (1993), "Fundamentals of a Zero-Point Energy Technology," <u>Proc. Int. Sym. on New Energy</u>, pp 201-217.

6. T.S. Kuhn (1970), <u>The Structure of Scientific Revolutions</u>, University of Chicago Press, Chicago.
   Shows that throughout history scientific paradigm shifts have been strongly resisted.

7. B. Rubick (1999), "The Perennial Challenge of Anomalies at the Frontiers of Science," *Infinite Energy*, vol 5, no 26, pp 34-41.
   Essay shows that paradigm violating experiments and anomalies are ferociously resisted by the scientific community past and present.

8. M.B. King (1997), "Charge Clusters: The Basis of Zero-Point Energy Inventions," *J. New Energy*, vol 2, no 2, pp 18-31; also *Infinite Energy*, vol 3, no 3-4, pp 96-102.
   Shows how many "free energy" inventions utilize (sometimes unwittingly) the phenomena of charge clusters, which may provide the coupling to the zero-point energy for their source of power.

9. K.R. Shoulders (1991), "Energy Conversion Using High Charge Density," U.S. Patent No. 5,018,180.
   Fundamental discovery on how to launch a micron size, negatively charged plasmoid called "Electrum Validum" (EV). An EV yields excess energy (over unity gain) whenever it hits the anode or travels down the axis of an hollow coil. The excess energy comes from the ZPE.

10. S.X. Jin, H. Fox (1996), "Characteristics of High-Density Charge Clusters: A Theoretical Model," *J. New Energy*, vol 1, no 4, pp 5-20.
A mathematical model of charged clusters (Shoulder's EV's) is presented that shows the stability is due to a helical vortex ring possessing an extraordinary poloidal circulation. In this nonrelativistic calculation, the poloidal filament would have to be thin. A spherical electron cluster is unstable and would tend to form into a toroid by a force balance relationship. The calculation shows that the energy density of the poloidal filament in a charge cluster is a hundred times higher than in a supernova explosion.

11. P.O. Johnson (1965), "Ball Lightning and Self Containing Electromagnetic Fields," *Am. J. Phys.*, vol 33, p 119.

12. I. Alexeff, M. Radar (1995), "Possible Precursors of Ball Lightning - Observation of Closed Loops in High-Voltage Discharges," *Fusion Tech.*, vol 27, pp 271-273.
Closed current loops were photographed during high voltage discharges. The loops enclose a magnetic field of very high energy density. They contract and quickly become compact force-free loops that superficially resemble spheres. In these toroidal geometries, the trapped internal magnetic field balances the external magnetic field to provide an almost force-free configuration. The bibliography cites numerous references on ball lightning.

13. R.W. Ziolkowski, M.K. Tippett (1991), "Collective effect in an electron plasma system catalyzed by a localized electromagnetic wave," *Phys. Rev. A*, vol 43, no 6, pp 3066-72.
Mathematical analysis of Shoulder's EV that includes a significant (vacuum polarization) displacement current term since the EV formation time is of the same order as the plasma frequency period. The resulting nonlinear Klein-Gordon equation contains vorticity terms and a term similar to a quantum mechanical potential which compensates for the repulsion. The system is solved by numerical methods for a stable, localized wave solution which matches the EV in size and charge density.

14. V. Nardi, W.H. Bostick, J. Feugeas, W. Prior (1980), "Internal Structure of Electron-Beam Filaments," *Phys. Rev. A*, vol 22, no 5, pp 2211-17.

15. D. Reed (1992), "Toward a Structural Model for the Fundamental Electrodynamic Fields of Nature," *Extraordinary Science*, vol IV, no 2, pp 22-33; ... (1993), "Evidence for the Screw Electromagnetic Field in Macro and Microscopic Reality," <u>Proc. Int. Sym. on New Energy</u>, pp 497-510; ... (1994), "Beltrami Topology as Archetypal Vortex," <u>Proc. Int. Sym. on New Energy,</u> pp 585-608; ...(1996), "The Beltrami Vector Field - The Key to Unlocking the Secrets of Vacuum Energy?" <u>Proc. Int. Sym. on New Energy,</u> pp 345-363.

16. G.A. Mesyats (1996), "Ecton Processes at the Cathode in a Vacuum Discharge," <u>Proc. 17th International Symposium on Discharges and Electrical Insulation in Vacuum,</u> pp 720-731.

Russian research is presented analyzing their discovery of charge clusters, called "ectons." Ectons often arise from micro explosions on the surface of the cathode, where surface imperfections such as micro protrusions, adsorbed gases, dielectric films and inclusions play an important role. The simplest way to initiate ectons is to cause an explosion of cathode micro protrusions under the action of field emission current. Experiments confirm micro protrusions jets can form from liquid or melting metal. The breakdown of thick dielectric films in their charging with ions also plays an important role in the initiation of ectons. A commonly used way to initiate an ecton is to induce a vacuum discharge over a dielectric in contact with a pointed, metal cathode. An ecton can readily be excited at a contaminated cathode with a low density plasma, but a clean cathode requires a high plasma density.

17. T.H. Moray, J.E. Moray (1978), <u>The Sea of Energy</u>, Cosray Research Institute, Salt Lake City.

18. P.N. Correa, A.N. Correa (1995), "Electromechanical Transduction of Plasma Pulses," U.S. Patent 5,416,391. "Energy Conversion System," U.S. Patent 5,449,989; ... (1996), "XS NRG™ Technology," *Infinite Energy*, vol 2, no 7, pp 18-38; ... (1997), "Metallographic & Excess Energy Density Studies of LGEN™ Cathodes Subject to a PAGD Regime in Vacuum," *Infinite Energy*, vol 3, no 17, pp 73-78.
Fundamental discovery that a plasma tube tuned to operate at the abnormal glow discharge region exhibits an over unity energy gain. The abnormal glow discharge is a glow plasma that surrounds the cathode just before a vacuum arc discharge (spark) that occurs when the tube is slowly charged with increasing voltage. The abnormal glow exhibits a negative resistance characteristic as the tube begins an arc discharge. The patents illustrate how to make an appropriate charging circuit to cycle the tube in the abnormal glow discharge regime, control the cycle frequency, avoid the losses of the vacuum arc discharge, and rectify the excess energy onto batteries.

19. P.M. Brown (1989), "Apparatus for Direct Conversion of Radioactive Decay Energy to Electrical Energy," U.S. Patent No. 4,835,433; ... (1987), "The Moray Device and the Hubbard Coil were Nuclear Batteries," *Magnets*, vol 2, no 3, pp 6-12; ... (1990), "Tesla Technology and Radioisotopic Energy Generation," <u>Proc. Int. Tesla Sym.</u>, Colorado Springs, chap 2, pp 85-92.
Brown created a five watt nuclear battery using a weak (one Curie) radioactive source, Krypton 85. Since the radioactive source could only provide at best five milliwatts, Brown created an anomalous self running energy device. Brown used an LC oscillator, where the radioactive material ionized a corona around the coil. If the circuit is tuned to the ion-acoustic resonance of the corona, then the ion-oscillations could couple a ZPE coherence directly to the circuit.

20. P.H. Matthey (1985), "The Swiss ML Converter - A Masterpiece of Craftsmanship and Electronic Engineering," in H.A. Nieper (ed.), <u>Revolution in Technology, Medicine and Society</u>, MIT Verlag, Odenburg.

21. D. Kelly, K. Stevens (1995), "The Swiss ML Converter," *Space Energy Journal*, vol VI, no 3, pp 13-31.

22. W.W. Hyde (1990), "Electrostatic Energy Field Power Generating System," U.S. Patent No. 4,897,592.

23. P. Graneau, P.N. Graneau (1985), "Electrodynamic Explosions in Liquids," *Appl. Phys. Lett.*, vol 46, no 5, pp 468-470.

24. P. Graneau (1997), "Extracting Intermolecular Bond Energy from Water," Proc. Fourth Int. Sym. on New Energy, pp 65-70; also *Infinite Energy*, vol 3, no 3-4, pp 92-95.
High speed photography reveals a plasma ball sitting in the accelerator muzzle of the water arc explosion experiments.

25. T. Ditmire, et al. (1997), "High Energy Explosion of Atomic Clusters: Transition from Molecular to Plasma Behavior," *Phys. Rev. Lett.*, vol 78, no 14, pp 2732-35.
Femtosecond laser pulse excitation of atomic noble gas clusters produce ion kinetic energies three orders of magnitude higher than expected.

26. J. Papp (1984), "Inert Gas Fuel, Fuel Preparation Apparatus and System for Extracting Useful Work from the Fuel," U.S. Patent No. 4,428,193; ...(1972), "Method and Means of Converting Atomic Energy into Utilizable Kinetic Energy," U.S. Patent No. 3,670,494.

27. E.V. Gray (1976), "Pulsed Capacitor Discharge Electric Engine," U.S. Patent No. 3,890,548.

28. G. Preparata (1991), "A New Look at Solid-State Fractures, Particle Emission and Cold Nuclear Fusion," *Il Nuovo Cimento*, vol 104 A, no 8, p 1259; ... (1990), "Quantum field theory of superradiance," in R. Cherubini, P. Dal Piaz, B. Minetti (editors), Problems of Fundamental Modern Physics, World Scientific, Singapore.

29. J. Bockris (1996), "The Complex Conditions Needed to Obtain Nuclear Heat from D-Pd Systems," *J. New Energy*, vol 1, no 3, pp 210-218.
The hypothesis is proposed that internal cracking of the cathode palladium (or nickel) is the needed triggering mechanism to obtain cold fusion or transmutation events. It explains why, even though it takes only three hours to load palladium rods to saturation, there can be delays of hundreds of hours before heat bursts occur. If the cracks should reach the surface, the deuterium fugacity is diminished and the reaction stops. Thin palladium nickel alloys or layers, as in Patterson's beads, allow the internal cracking to occur quickly giving reliable and repeatable results.

30. K. Shoulders, S. Shoulders (1996), "Observations on the Role of Charge Clusters in Nuclear Cluster Reactions," *J. New Energy*, vol 1, no 3, pp 111-121.
Experimental evidence in the form of micrographs and X-ray microanalysis is presented suggesting that nuclear charge clusters, (micron sized plasmoids containing 10^{11} electrons and 10^6 protons or deuterons) can

accelerate into lattice nuclei with sufficient kinetic energy to overcome the Coulomb barrier and trigger transmutation events. The hypothesis to explain cold fusion is proposed where electrolytic loading of palladium or nickel causes cracking and fractoemission of the charge clusters.

31. J. Patterson (1996), U.S. Patent 5,494,559. Also J. Rothwell (1995), "Highlights of the Fifth International Conference on Cold Fusion," *Infinite Energy*, vol 1, no 2, pp 8-18.
The first U.S. "cold fusion" patent with the claim of excess energy (2000%) granted. The Patterson cell contains hundreds of sub millimeter beads made in a electroplating process of depositing multiple, alternating, thin layers of very pure nickel and palladium. It runs using a light water electrolyte. It is considered the best of the electrolytic cold fusion type of experiments with complete repeatability and a record gain of over 1000 in one demonstration.

32. F. Sweet, T.E. Bearden (1991), "Utilizing Scalar Electromagnetics to Tap Vacuum Energy," Proc. 26th IECEC, vol 4, pp 370-375; ... (1988), "Nothing is Something: The Theory and Operation of a Phase-Conjugate Vacuum Triode;" ... (1989), private communication.

33. W. Rosenthal, D. Watson (1999), private communications.

34. M.B. King (1994), "Vacuum Energy Vortices," Proc. Int. Sym. on New Energy, pp 257-269.
Discusses force-free vortices. Suggests that counter-rotating plasmas can induce a ZPE coherence akin to a large scale "pair production."

35. M.B. King (1998), "Vortex Filaments, Torsion Fields and the Zero-Point Energy," *J. New Energy*, vol 3, no 2-3, pp 106-116.
Overview of Russian research on torsion fields suggest plasma vortex action can induce large vacuum torsion fields in the ZPE.

36. A.E. Akimov (1998), "Heuristic Discussion of the Problem of Finding Long Range Interactions, EGS-Concepts," *J. New Energy*, vol 2, no 3-4, pp 55-80.

37. A.E. Akimov, G.I. Shipov (1997), "Torsion Fields and Their Experimental Manifestations," *J. New Energy*, vol 2, no 2, pp 67-84.

Transforming the Planet with a Zero-Point Energy Experiment

Transforming the Planet with a Zero-Point Energy Experiment

Dual Vortex Forms:
The Key to a Large Zero-Point
Energy Coherence

October 2000

Abstract

A hypothesis is suggested that the vacuum's zero-point energy can self-organize into vortical forms at any size scale. The vacuum naturally manifests vortical pairs of opposite chirality (left handed and right handed helicity), and it is the chirality that gives each vortex form its charge polarity. At the fermi scale the vortices are an electron/positron pair. At a larger scale the vortices become a pair of "quasi" charges. Shoulders' electrum validum, an anomalously energetic, micron size, plasma-like formation might be an example of quasi charge. Inventions that make properly shaped, dual vortex, plasma forms might produce quasi charge from the zero-point energy and output excessive power.

Introduction

The hypothesis that elementary charge is a vortical flow in a fluid aether was popular in the 19th century. Here the model will be reexamined where the zero-point energy (ZPE) plays the role of the aether. The model implies the existence of "quasi" charge, a vortical form the same shape as the presumed vortex making an elementary charge, but at a larger size scale. Shoulders' discovery of the electrum validum (EV) might be experimental evidence of micron sized quasi charge. Once the exact vortical shape is known, technological means could induce a pair of centimeter scale quasi charges, and these might generate excessive energy directly from the fabric of space.

There exists controversy in the scientific community regarding the nature of pure, empty space. Undergraduate textbooks in physics explain that the vacuum is a void because Einstein's theory of relativity requires it. The hypothetical aether suggested in the 19th century to support the propagation of light waves was deemed an unnecessary construct since the principle of relativity required all inertial observers (those moving in free space at constant velocity) must observe the same laws of physics. A simplistic, static aether would manifest an aether wind proportional to the observer's velocity, and such a wind was not detected in the famous Michelson-Morley experiments. Moreover, the elegance of the mathematical derivation of Maxwell's equations from special relativity established Lorentz invariance as a fundamental principle that must be fulfilled by all subsequent foundational theories in physics. Since a complex Lorentzian aether model seems contrived, Occam's razor is generally used to rule it out. Today it is obvious to most of the scientific community (who are generally not physicists) that empty space is a void.

However, modern quantum physicists have a different model for the fabric of space. Quantum mechanics has a term in its equations that describes an underlying jitter to all systems known as the zero-point energy or zero-point fluctuations (ZPF). Zero-point refers to absolute zero degrees Kelvin and implies the energy is not propagating like heat or photon radiation, but rather it is energetic, electric field fluctuations that are imbedded in "raw" empty space. The ZPE is the basis of quantum electrodynamics [1] and has a density sufficient to manifest short-lived (virtual) pairs of elementary particles. Its characteristic spectrum is Lorentz invariant allowing its physical interaction to be modeled in a covariant fashion consistent with the theory of relativity. (All inertial observers see the same ZPE spectrum). There is a point of view, well represented in the physics journals, that suggests that quantum physics is a direct result of matters' interaction with the ZPE [2]. Moreover, it is proposed that the ZPE is really the foundation of physical world, and it could in principle be used to explain elementary particles [3], atomic stability [4], inertia [5], and gravity [6]. There is a severe dichotomy between the view of the quantum physicists and the (non physicist) majority of the scientific community regarding empty space. Is it a plenum or void?

The question regarding what is the true nature of empty space manifests as a "paradigm war" [7] with different "camps" expressing their opinion. Those that believe the vacuum is a void say that the ZPE is just a mathematical fiction. Others say that the vacuum fluctuations are present but are so small that they can effectively be ignored. Another group concedes that the vacuum fluctuations could be large, but entropy makes coherence unlikely because the fluctuations are random and chaotic. This argument can be refuted by chaos theory where the conditions for system self-organization [8] (open flux, nonlinear, far from equilibrium) can allow its coherence. Perhaps the most powerful theory of the

ZPE is Wheeler's geometrodynamics [9] where the vacuum fluctuations arise from electric flux from a physical fourth dimension. Here electric flux enters and immediately leaves our three dimensional space through Planck length (10^{-33} cm) diameter channels called "wormholes" that contain extraordinary energy densities, 10^{94} g/cm^3. Wheeler describes the resulting chaotic action manifested in 3-space as the "quantum foam." It exhibits characteristics similar to turbulent plasma with opportunities for self-organization, and these typically manifest as pair production of short lived, elementary particles. Inducing even a slight coherence on a macroscopic scale would offer the potential for a limitless energy source. The stakes for humanity in this paradigm war are huge: The majority view is "don't even look" for space is empty. On the other hand a new abundant energy source could be awaiting our discovery, if we only look.

Implications of Hyperspace

If the existence of a physical hyperspace appears outlandish, there seems to be an even bigger paradigm shift in store for humanity arising from the conclusions of physicists exploring the foundational issues of quantum mechanics. There is both theoretical and experimental recognition of instantaneous connectivity between distant elementary particles that are "quantum entangled" due to a previous interaction [10]. The foundational physicists are suggesting either a non-local connecting mechanism exists outside of space-time [11] or there exists an infinitude of "parallel" three-dimensional universes [12]. Some even suggest that time itself does not really exist [13], but rather it is a partial perception of our limited consciousness slicing a fourth spatial dimension. It seems that the foundational physicists are heading toward the philosophical model of Ouspensky's Tertium Organum [14], a work first published in 1912 (twenty years prior to the discovery of quantum mechanics). Ouspensky emphasized that one extra spatial dimension is all that is required to embed an infinite number of three-dimensional universes. The notion that space might contain extractable, useful energy is a small paradigm shift when compared to the implications of non-local connections and the illusionary existence of time.

Shifting the Paradigm

A paradigm shift in science cannot be caused by mere theoretical discussions. Only an experiment will do, and even if such an experiment is created, it is typically ignored or ridiculed by the established scientific community [15]. For example in the early1900's the scientific press scorned the Wright brothers even after they flew 26 miles [16]. Today the cold fusion experiments are derided despite repeatable evidence of excess heat and nuclear transmutation [17]. Also if a paradigm shifting invention threatens entrenched economic interest, it would likely come under severe suppression [18]. The way to overcome such resistance is to have massive replication of an invention by the community at large [19]. To have a lasting impact, an invention that demonstrates the ZPE as an energy source must be simple and robust so that it is easily communicated and replicated. It must also be self-running so there is no controversy regarding energy measurement errors. If the vacuum contains a plenum of energy, there should be principles that allow such an invention to be discovered.

Zero-Point Energy Engineering

Fortunately there are principles in the scientific literature that could be applied to engineer the vacuum energy. In 1977 Prigogine won the Nobel Prize in chemistry for identifying in general system theory terms the conditions required for a system to evolve from chaos to self-organization. The system must be nonlinear, far from equilibrium, and have a flux (of energy or of matter) through it. Wheeler's geometrodynamics model of the vacuum fulfills these conditions. Similarly a ZPE device would likely have a nonlinear coupling component, which is typically driven off of equilibrium by an abrupt electrical discharge. Moreover, the coupling component's constituent particles should maximally influence the surrounding ZPE to channel its flux into the system. Which particles maximize the interaction?

The descriptions from quantum electrodynamics of vacuum polarization of the elementary particles can be used to guide how to maximize the interaction with the ZPE (Figure 1). Electrons, especially those in ordinary conductors, are described as a smeared electron cloud that is essentially in thermodynamic equilibrium with the ZPE [20]. Thus there are little net vacuum energy effects with ordinary electronic circuitry. On the other hand atomic nuclei exhibit steep, converging lines of vacuum polarization, which makes them ideal activators for coherent vacuum effects. The cohering effects are observed in collision experiments [21] as well as plasma ion-acoustic resonance [22] where large energetic anomalies are noticed. The motion of charge nuclei can play a significant role in an invention that coheres the zero-point energy.

Quantum electrodynamics also describes the principles underlying pair production of charged particles directly from the vacuum. Spin and charge are both conserved. Conservation of charge is easy to model in geometrodynamics.

Electric flux entering our 3-space manifests one charge polarity; the flux exiting manifests the other. Charge conservation is simply continuity of flux. Spin is tougher to model. The spin 1/2 particles exhibit a two-valued state where a 360-degree rotation does not restore the original state; a 720-degree rotation does [23]. Thus an elementary particle is not a simple three-dimensional form. Russian torsion field physicists [24] have described the vacuum as giving rise to "non-orientable topological entities" in a form like a Klein bottle, the 4-space analog to the Mobius strip. Its 3-space projection appears like a vortex ring, but its rotation includes a fourth dimensional, "inside-out" component. Unfortunately exact details are beyond the scope of today's scientific literature. The details are important for they could guide what to do at the macroscopic level to induce large zero-point energy coherence. The underlying hypothesis is that whatever nature does in the vacuum at the elementary particle scale to manifest electron-positron pairs, can also be induced at larger scales of self-organization. The quantum foam might support creating "quasi" charge pairs at any scale, and the act of doing so at the macroscopic level might manifest spectacular energetic effects.

Electrum Validum

The discovery of macroscopic quasi charge formation appears to have already been made by Shoulders [25]. An abrupt electric discharge from a pointed cathode near a dielectric surface creates a micron size plasmoid with a net equivalent charge on the order of 10^{23} electrons. Shoulders named this charge formation "electrum validum" (EV). Most investigators have assumed the EV is comprised of a cluster of electrons, and then tried to explain how such a cluster could be stable in the face of Coulomb repulsion. Attempts have been made to model it classically as a vortex ring [26], nonlinearly as a soliton using vacuum polarization to provide a compensating attraction [27], classically using the polarized dielectric to provide an attractive force [28], and recently Shoulders [29] suggested the EV might alter the local permittivity of space to trap the electrons via surface tension within a permittivity transition "bubble." Here will be offered an alternative hypothesis: The EV is not a collection of electrons but rather is a manifestation of a pair production of two macroscopic quasi charges from the vacuum. The negative quasi charge is above the dielectric surface; its positive counterpart is below. The pair can be crudely modeled from a torsion field perspective as two, subtle, counter-rotating vortices in the quantum foam stabilized by the presence of the dielectric surface. Moreover, the hypothesis suggests that the excessive energy exhibited when the EV strikes a conductor arises from the quasi charge, pair annihilation. The quasi pair is a direct coherence of the ZPE, and in principle can form at any size scale.

The creation of counter-rotating vortical overlap might yield subtle formations, which later could be triggered into an active energetic state. Shoulders' empirical studies of the EV phenomena might lend support for subtle, yet potentially energetic forms. He observed that an EV sometimes could turn invisible (black) and while in this state manifest no net electric charge. It could later be electrically stimulated to once again ionize its local environment, emit light and manifest a huge electric charge. Also multiple EV's could be organized into complex chains and rings that stay together even while in the black state. Surprisingly, Shoulders has occasionally detected positive EV's in experiments that predominantly produce the negative variety, and they exhibit the e/m ratio of positrons even though their decay characteristics do not manifest gamma rays from electron/positron annihilations. The positive EV is perhaps the strongest evidence for the conjecture of macroscopic quasi charge formation.

The EV's must stay near a dielectric boundary to remain stable. An EV launched into a vacuum travels a zigzag path and quickly explodes. An EV pair launched into a vacuum exhibits a double helix spiral path before it explodes. The dielectric appears to be supporting appropriate counter charge to stabilize the EV's existence. The black EV implies the counter charge and the original charge can overlap sufficiently to manifest charge neutrality without annihilation. Could the permittivity difference across the dielectric boundary prevent the quasi pair annihilation? Could a black EV be stimulated to exhibit a positive charge even if it originally started out as negative? Clearly Shoulders has discovered a complex form that probes the essence of charge and its relationship to the vacuum.

Torsion Fields

Perhaps the best theoretical support for the notion of quasi charge formation comes from the Russian research on torsion fields [30]. Here the concept of counter-rotation is fundamental, and it appears at the microscopic foundation of the physical vacuum in the form of a "phyton," a hypothetical sub quantum particle, scaled at the Planck length, which exhibits an inherent dual rotation. (Note that a spatial fourth dimension appears implicit in order to have one rotation "overlap" the other.) The phyton is able to separate its counter-rotating parts in different directions, and the various polarizations of a lattice of phytons can manifest electric fields, torsion fields, and gravitational fields. The theory supports large-scale organization in the vacuum based on spin and chirality (left handed and right handed helical forms). The vacuum forms may separate dramatically to manifest strong fields and pair production, or the forms may overlap to manifest subtle fields or apparently nothing. Note that a persistent organization in the overlap state implies a separation in the fourth dimension to avoid collision in the counter-rotating flows. Russell [31] likewise stressed the theme of overlapping counter-rotation in his esoteric treatise, The Universal One, with numerous archetypal diagrams of opposing vortices to describe nature at all levels, from elementary particles to galaxies.

Inventions

Counter-rotation appears in some "free energy" inventions. The Swiss ML Converter [32] counter spins two clear acrylic disks, which excite a colorful corona via high voltage electrostatics in the gap between them. The device reportedly outputs kilowatts of power while self-running. Another reported invention [33] stimulated a flat coil with surrounding counter-rotating copper disks that impressed bucking fields from peripherally mounted magnets. The apparatus was placed in the field above a Tesla coil (with the spin axis aligned with the coil axis). The device reportedly produced excess energy.

In similar fashion a resonantly energized, Tesla pancake coil sandwiched between counter-rotating dielectric disks which hold the magnets might induce a large output. The counter-rotating, bucking magnetic fields can stimulate the corona surrounding the pancake coil. If the resulting corona motion induces dual vortices in the ZPE, the device could manifest energetic and gravitational anomalies. Counter-rotation might be a potent engineering principle for activating a significant ZPE organization.

Dual Vortex

Perhaps a clue as to the nature of charge as an organized structure in the quantum foam can be inferred from the esoteric work of Leadbeater [34], who modeled charge as a double vortex of flowing aether, which he named the "anu" (Figure 2). Charge of opposite polarity has opposite chirality. Leadbeater suggested the positive charge circulation twists flux from a fourth dimension into our 3-space to become a source while the negative charge circulation induces a corresponding sink. The illustrations show a centripetal inner vortex surrounded by an outer circulation connected by a Mobius twist. The diagrams are similar to the implosion vortex that Schauberger [35] described to be the critical form for water flow to manifest an anomalously energetic effect. Schauberger claimed to measure a propulsive gain as well as observe a bluish glow near the tip of the implosion vortex in his water experiments. Note the bluish glow is similar to that observed in sonoluminescence [36], which has been attributed to a ZPE interaction [37]. The suggested hypothesis is this: Circulating charged matter or plasma in a dual vortex flow like the anu might manifest a corresponding resonant form in the ZPE.

As a supporting example, dual vortex action appears to be exhibited in the gasoline engine, fuel manifold invented by Pantone [38] (Figure 3). Here an intake pipe guides the downward flowing fuel into a vortex due to a conically tapered, steel rod mounted flexibly and centrally along the pipe's axis. A concentric exhaust manifold surrounds the intake pipe, which helps to vaporize the fuel, and it supports a vortical upward exhaust flow. Since the gap between the central steel rod and the intake pipe wall is so small (0.05 inch), the fuel vapor would partially ionize due to friction. If the ionic vortex induced a resonant formation in the ZPE, it might explain some of Pantone's claims associated with the invention such as pollution free exhaust, ability to run with crude fuels or non fuels (e.g. water) and observation of glowing plasma-like activity within the intake pipe. Pantone's invention has intrigued many investigators, and it might be an example of vortex action inducing coherence in the zero-point energy.

The hypothesis that a dual vortex induces a large ZPE coherence could be investigated by constructing a vortex globe designed to support the anu shaped flow (Figure 4). Here an outer heart shaped, glass (or clear plastic) chamber surrounds a transparent inner pipe. The inner pipe has a funnel at the top, and optionally a conically tapered rod can sit loosely in the pipe similar to Pantone's manifold. The rod could optionally have spiral grooves to help shape the vortex. Alternatively, fluting the inner pipe could shape the vortex instead of using a central rod. The vortex can be started by blowing air circumferentially around the globe's interior. Feed tubes inserted tangentially near the top could direct the circumferential airflow. Alternatively, a spinning propeller or magnetic stirrer (sitting inside, near the bottom) can create a vortex within a sealed globe. Large energetic effects might be achieved by circulating plasma or charged particles instead of just air. If some fine sand were put into the globe, the vortex motion would tend to ionize and charge the sand particles via electrostatic friction. The electrostatic activity would naturally produce corona, which would be expected to get brighter if ZPE coupling occurs. Also a voltage gradient might be induced between the top and bottom of the globe. If such a voltage occurs, experiments to detect "cold current" could be tried by connecting thin wires from the top and bottom of the globe to a lamp. The hypothesis is that cold current arises from a vortical, vacuum polarization displacement current surrounding the wire rather than standard electron conduction. Moray [39], Bedini [40] and Sweet [41] observed cold currents in their inventions that produced excess energy. It might be a definitive indicator that the ZPE is being activated as the energy source.

The largest effects might occur using two globes whose vortices are spinning oppositely. Here conservation of angular momentum and conservation of charge fulfill the dynamics of vacuum pair production. If the anu vortex mimics a macroscopic form of quasi charge, creating a balanced pair might tend to align naturally with how the physical vacuum behaves. If the positive anu sources electric flux from a fourth dimension, and if the negative anu acts as a corresponding sink, then experiments of connecting an electrical load between the globes might yield interesting results (Figure 5). Further experiments by completing the circuit loop (from the bottom to the top) might augment the output energy. Mimicking pair production on a macroscopic scale could be an effective way to tap significant zero-point energy.

Summary: Principles for ZPE Coherence

The principles for cohering the vacuum energy are motivated by Wheeler's quantum foam model of the physical vacuum where an energetic turbulent substrate can yield macroscopic self-organized forms in a manner similar to elementary particle pair production in quantum electrodynamics. The following principles could inspire the creation of many "free energy" inventions:

1. Drive the system into a far from equilibrium state, typically with an abrupt electrical discharge. The far from equilibrium state is required for self-organization.
2. Work with a nonlinear system. This is likewise required for self-organization. A component containing plasma or corona can make the system nonlinear.
3. Accelerate those particles that maximize their vacuum polarization interaction with the ZPE. These are typically nuclei or ions.
4. Surround the ZPE coupling component with abrupt electromagnetic field compression and release. Such activity can help twist the fourth dimensional ZPE flux into our 3-space [42]. Pulsing

caduceus wound coils, bifilar coils, or conductive Mobius strips can create the bucking field conditions.

5. Use counter-rotation for spinning systems. Vacuum dynamics conserve angular momentum.

6. Create vortical forms. These include vortex filaments, vortex rings and dual vortex forms like the anu. Mimic on a large scale a vortical model for elementary charge.

7. Mimic pair production. Since vacuum dynamics conserve charge, vortical flows intended to make quasi charge should be created in balanced, opposing pairs.

8. Use high voltage to stress and polarize the vacuum. Electrostatic or frictional means can produce very high voltages. Tesla coils can do likewise. Abrupt, high voltage discharges can create anomalously energetic forms (like EV's). Couple their energy to the system.

From these principles a wide variety of zero-point energy devices might be invented. If just one self-running device were shared with the scientific community for worldwide replication, a paradigm shift would occur. The shift would give birth to a new industry that would produce a plethora of improved inventions and yield a new technological age for mankind.

Acknowledgements

The author wishes to thank Neil Boyd and Ken Shoulders for helpful discussions.

Dual Vortex Forms: The Key to a Large Zero-Point Energy Coherence

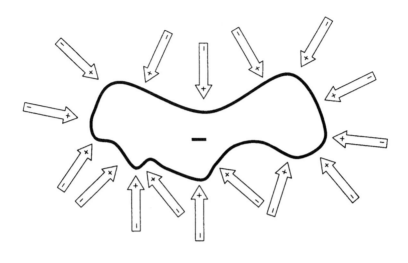

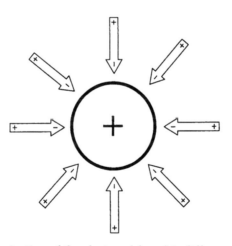

Fig 1 Vacuum polarization of the electron (above) is different than that for the atomic nucleus (below). Conduction band electrons behave like a smeared charged cloud in thermodynamic equilibrium with the vacuum fluctuations. In contrast, nuclei exhibit steep, converging lines of vacuum polarization, and thus their appropriate vortex motion of plasma ions might induce a large-scale coherence.

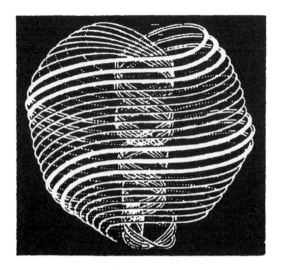

POSITIVE

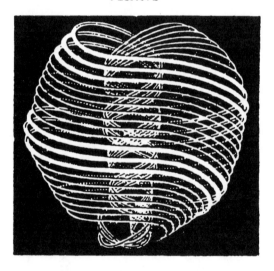

NEGATIVE

Fig 2 Leadbeater's "anu" is a dual vortex aether model for charge. Spin direction determines the charge's polarity. Positive polarity twists flux from the fourth dimension into our 3-space to manifest a source; negative polarity manifests a corresponding sink. The proposed hypothesis is that the vacuum energy can naturally self-organize into the anu vortex at large size scales to become a "quasi" charge. Circulating plasma in the anu vortex shape might induce such a quasi charge and yield a large net energy gain.

186

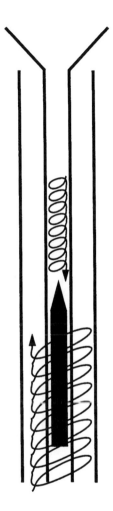

Fig 3 Cross section of Pantone's fuel manifold shows how it can support a dual vortex. Inner pipe guides fuel down to a tapered steel rod which induces ionization and vorticity in the narrow gap. Exhaust flowing vortically upward in the surrounding outer pipe preheats the fuel.

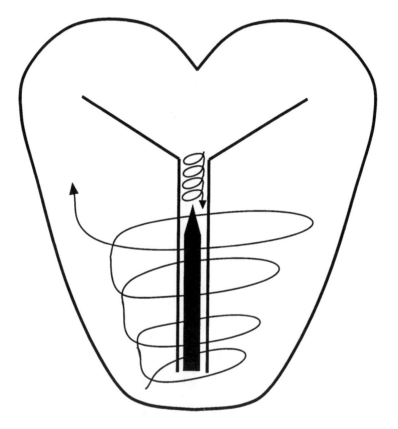

Fig 4 Cross section of a vortex globe designed to make the "anu" shaped, dual vortex. Funnel guides flow to the inner pipe where a tapered rod induces a spiral. A magnetic stirrer (not shown) can act as a propeller to control the spin in a sealed globe. Some fine sand mixed into an air vortex would electrostatically charge, and its circulation might mimic a plasma vortex. If the vortex induced a large scale quasi charge in the zero-point energy, the globe would exhibit a positive or negative polarity depending on the direction of spin.

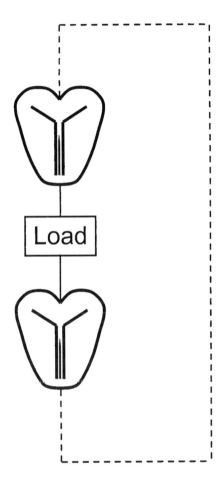

Fig 5 Experiments with two "anu" vortex globes spinning oppositely might exhibit cold current where thin wires guide vortical, vacuum polarization displacement current. If the vortex globes produce a pair of large quasi charges, which act like a flux source and sink, a complete circuit loop (dashed line) might not be required.

Dual Vortex Forms: The Key to a Large Zero-Point Energy Coherence

References

1. R.P. Feynman, QED The Strange Theory of Light and Matter, Princeton Univ. Press, Princeton, NJ, 1985; ... "Space-Time Approach to Quantum Electrodynamics," *Phys. Rev.*, vol 76, p 769 (1949).

2. T.H. Boyer, "Random Electrodynamics: The theory of classical electrodynamics with classical electromagnetic zero-point radiation," *Phys. Rev. D*, vol 11(4), pp 790-808 (1975); ... "Derivation of Blackbody Radiation Spectrum without Quantum Assumptions," *Phys. Rev.*, vol 182(5), pp 1374-83 (1969).

3. R.C. Jennison, "Relativistic Phase-Locked Cavities as Particle Models," *J. Phys. A Math Gen.*, vol VII(8), pp 1525-33 (1978); ... "A New Classical Relativistic Model of the Electron," *Phys. Lett. A*, vol 141(8/9), pp 377-382 (1989).

4. H.E. Puthoff, "Ground state of hydrogen as a zero-point fluctuation determined state," *Phys. Rev. D,* vol 35(10), pp 3266-69 (1987).

5. B. Haisch, A. Rueda, H.E. Puthoff, "Inertia as a zero-point field Lorentz force," *Phys. Rev. A*, vol 49(2), pp 678-694 (1994).

6. H.E. Puthoff, "Gravity as a zero-point fluctuation force," *Phys. Rev. A*, vol 39(5), pp 2333-42 (1989).

7. M.B. Woodhouse, Paradigm Wars, Frog Ltd., Berkeley, CA, 1996.
    Discussion of world views and paradigm shifts across many fields including system holism, system self-organization, new physics, non-locality, higher dimensions, energy, time, consciousness, transpersonal psychology, religion, Perennial philosophy, mysticism, health, education, environment and extraterrestrial contact.

8. I. Prigogine, G. Nicolis (1977), Self Organization in Nonequilibrium Systems, Wiley, NY, 1977; I. Prigogine, I. Stengers, Order Out of Chaos, Bantam Books, NY, 1984.

9. J.A. Wheeler, Geometrodynamics, Academic Press, NY, 1962.

10. A. Einstein, B. Podolsky, N. Rosen, "Can Quantum Mechanical Description of Physical Reality be Considered Complete?" Phys. Rev., vol 47, p 777 (1935).
    The original paper, which showed that in quantum mechanics, particles born from the same quantum event remain strongly correlated even when separated. Experiments in the 1980's confirmed it. The essence of the paradigm shift from classical physics is that the elementary particles cannot be "locally" modeled. See

Dual Vortex Forms: The Key to a Large Zero-Point Energy Coherence

G. Zukav, The Dancing Wu Li Masters, Bantam Books, NY, 1979, for a thorough discussion.

11. J. Gribbin, Schrodinger's Kittens and the Search for Reality, Back Bay Books, NY, 1995.
    Overview of the foundational issues (non-local connectivity) and interpretations of quantum mechanics including the Copenhagen interpretation, many-worlds interpretation, and the transactional interpretation where distant quantum entangled objects can be instantaneously connected across both space and time.

12. D. Deutsch, The Fabric of Reality, Penguin Books, NY, 1997.
    The leading proponent of the many-worlds interpretation discusses quantum mechanics, epistemology, theory of computation, and the theory of evolution yielding a view of reality in which the past, present and future simultaneously exist.

13. J. Barbour, The End of Time, Oxford University Press, NY, 2000.
    A physicist proposal that the next revolution in physics will recognize the passage of time does not physically exist but is really an illusion of consciousness. Barbour shows how a timeless "best fit" hyper-surface algorithm is mathematically equivalent to the standard Hamiltonian time-evolving dynamics for all descriptions of physics including Newtonian mechanics, general relativity and quantum mechanics. It yields a many-worlds model without the passage of time.

14. P.D. Ouspensky, Tertium Organum, Vintage Books, NY, 1970.
    Philosophical treatise written in 1912 proposes that the perception of time is a result of the mind's partial perception of a higher physical dimension. For human consciousness, a spatial fourth dimension exists, and our partial perception of it is experienced as the passage of time. Since one extra physical dimension is sufficient to contain an infinite number of three-dimensional universes, this model sets the foundation for a timeless many-worlds interpretation of reality. It appears to be a model to which many of today's quantum physicists/philosophers are heading, and it is remarkable that it was published twenty years before the discovery of quantum mechanics.

15. T.S. Kuhn, The Structure of Scientific Revolutions, University of Chicago Press, Chicago, 1970.
    Shows that throughout history scientific paradigm shifts have been strongly resisted.

16. J. Rothwell, "The Wright Brothers and Cold Fusion," *Infinite Energy*, vol 2(9), pp 37-43 (1996).

Dual Vortex Forms: The Key to a Large Zero-Point Energy Coherence

17. B. Rubick, "The Perennial Challenge of Anomalies at the Frontiers of Science," *Infinite Energy*, vol 5, issue 26, pp 34-41 (1999).
    Essay shows that paradigm violating experiments and anomalies are ferociously resisted by the scientific community in both the past and the present.

18. J. Eisen, Suppressed Inventions and Other Discoveries, Avery Publishing Group, Garden City Park, NY, 1999.

19. M.B. King, "Transforming the Planet with a Zero-Point Energy Experiment," *J. New Energy*, vol 4(2), pp 105-114 (1999). Also *Infinite Energy*, vol 6, issue 34, pp 51-55 (2000).

20. I.R. Senitzky, "Radiation Reaction and Vacuum Field Effects in Heisenberg-Picture Quantum Electrodynamics," *Phys. Rev. Lett.,* vol 31(15), p 955 (1973).

21. L.S. Celenza, V.K. Mishra, C.M. Shakin, K.F. Liu, "Exotic States in QED," *Phys. Rev. Lett.*, vol 57(1), p 55 (1986); D.G. Caldi, A. Chodos, "Narrow e^+e^- peaks in heavy-ion collisions and a possible new phase of QED," *Phys. Rev. D*, vol 36(9), p 2876 (1987); Y. Jack Ng, Y. Kikuchi (1987), "Narrow e^+e^- peaks in heavy-ion collisions as possible evidence of a confining phase of QED," *Phys. Rev. D*, vol 36(9), p 2880 (1987); L.S. Celenza, C.R. Ji, C.M. Shakin, "Nontopological solitons in strongly coupled QED," *Phys. Rev. D*, vol 36(7), pp 2144-48 (1987).

22. Yu G. Kalinin, et al., "Observation of Plasma Noise During Turbulent Heating," *Sov. Phys. Dokl.,* vol 14(11), p 1074 (1970); H. Iguchi, "Initial State of Turbulent Heating of Plasmas," *J. Phys. Soc. Jpn.,* vol 45(4), p 1364 (1978); A. Hirose, "Fluctuation Measurements in a Toroidal Turbulent Heating Device," *Phys. Can.*, vol 29(24), p 14 (1974).

23. C. Misner, K. Thorne, J. Wheeler, Gravitation, W.H. Freeman, NY, 1970.
    Thorough text on general relativity and differential forms. Chapter 43 discusses the zero-point energy and geometrodynamics. Chapter 44 discusses the problems of modeling actual charge and physics beyond the singularity of gravitational collapse.

24. I.M Shakhparonov, "Kozyrev-Dirac Emanation Methods of Detecting and Interaction with Matter," *J. New Energy*, vol 2(3-4), pp 40-45 (1998).
    Overviews concepts and experiments of creating "non-orientable" topological structures using conductive Mobius band circuit elements. Such topological structures have a hyperspatial nature and cannot be oriented in three-dimensional space. The projection into three space manifest phenomena akin to ball lightning. Magnetic monopoles are likewise of this nature, and experiments are referenced where beams of such are launched, detected and shown to exhibit alterations in gravity as well as the pace of time.

25. K.R. Shoulders, "Energy Conversion Using High Charge Density," U.S. Patent No. 5,018,180 (1991).

Fundamental discovery on how to launch a micron size, negatively charged plasmoid called "Electrum Validum" (EV). An EV yields excess energy (over unity gain) whenever it hits the anode or travels down the axis of an hollow coil. The excess energy comes from the ZPE.

26. S.X. Jin, H. Fox, "Characteristics of High-Density Charge Clusters: A Theoretical Model," *J. New Energy*, vol 1(4), pp 5-20 (1997).

A mathematical model of charged clusters (Shoulder's EV's) is presented that shows the stability is due to a helical vortex ring possessing an extraordinary poloidal circulation. In this nonrelativistic calculation, the poloidal filament would have to be thin. A spherical electron cluster is unstable and would tend to form into a toroid by a force balance relationship. The calculation shows that the energy density of a charge cluster is a hundred times higher than in a supernova explosion.

27. R.W. Ziolkowski, M.K. Tippett, "Collective effect in an electron plasma system catalyzed by a localized electromagnetic wave," *Phys. Rev. A*, vol 43(6), pp 3066-72 (1991).

Mathematical analysis of Shoulder's EV that includes a significant (vacuum polarization) displacement current term since the EV formation time is of the same order as the plasma frequency period. The resulting nonlinear Klein-Gordon equation contains vorticity terms and a term similar to a quantum mechanical potential, which compensates for the repulsion. The system is solved by numerical methods for a stable, localized wave solution, which matches the EV in size and charge density.

28. P. Beckmann, "Electron Clusters," *Galilean Electrodynamics*, vol 1(5), pp 55-58 (1990).

Explains how Shoulders' EV can be stabilized by a polarization interaction with the adjacent dielectric.

29. K.R. Shoulders, "Permittivity Transitions," *J. New Energy*, vol 5(2), pp 121-137 (2000).

Observations from electrum validum (EV) experiments include coupled EV pairs traveling in a double helix, positive EV's that exhibit a charge to mass ratio like the positron, and black (invisible) EV's, which do not manifest any interactions until they are stimulated. Shoulders suggests the EV modifies the permittivity of space to create self stability.

30. A.E. Akimov, "Heuristic Discussion of the Problem of Finding Long Range Interactions, EGS-Concepts," *J. New Energy*, vol 2(3-4), pp 55-80 (1998).

Dual Vortex Forms: The Key to a Large Zero-Point Energy Coherence

Overview of the torsion field research (predominantly in the Soviet Union) includes 177 references. Akimov models the vacuum as a lattice of "phytons," counter-rotating, charged entities sized at the Planck length (10^{-33} cm). Each phyton can polarize in three different ways to manifest 1) electric fields via charge polarization, 2) gravitational fields via oscillating, longitudinal spin polarization, and 3) torsion fields via transverse spin polarization. Torsion fields can arise from four sources: 1) physical classical spin, 2) the spin of the elementary particles comprising an object, 3) electromagnetic fields, and 4) the geometric form of the object. The spin polarized phyton lattice can also retain a temporary residual torsion field image of a (long standing) stationary object after it is moved.

31. W. Russell, The Universal One, University of Science and Philosophy, Waynesboro, Virginia, 1974.

32. P.H. Matthey, "The Swiss ML Converter - A Masterpiece of Craftsmanship and Electronic Engineering," in H.A. Nieper (ed.), Revolution in Technology, Medicine and Society, MIT Verlag, Odenburg, 1985.

33. R. Meyers, G. Perry, "Inter-dimensional Power System," *Extraordinary Science*, vol 5, issue 2, p 36 (1993).

34. C.W. Leadbeater, A. Besant, Occult Chemistry, reprinted by Kessington Publishing, Kila, Montana.
    Clairvoyant description of matter made in the years from 1895 through 1933. The fundamental constituent of matter was a dual vortex form named the "anu," whose chirality determines the charge polarity.

35. O. Alexandersson, Living Water: Viktor Schauberger and the Secrets of Natural Energy, Gateway Books, Bath, UK, 1990. Also B. Frokjaer-Jensen, "The Scandinavian Research Organization and the Implosion Theory (Viktor Schauberger)," Proc. First International Symposium on Nonconventional Energy Technology, Toronto, 1981, pp 78-96.

36. B.P. Barber, S.J. Putterman, "Observation of synchronous picosecond sonoluminescence," *Nature*, vol 353, pp 318-320 (1991); ... "Light Scattering Measurements of the Repetitive Supersonic Implosion of a Sonoluminescing Bubble," *Phys. Rev. Lett.*, vol 69, pp 3839-42 (1992).

37. C. Eberlein, "Sonoluminescence as Quantum Vacuum Radiation," *Phys. Rev. Lett.*, vol 76, pp 3842-45 (1996); Ö "Theory of quantum radiation observed as sonoluminescence," *Phys.Rev. A*, vol 53, pp 2772-87 (1996).

38. P. Pantone, "Fuel Pretreater Apparatus and Method," U.S. Patent No. 5,794,601 (1998).

Dual Vortex Forms: The Key to a Large Zero-Point Energy Coherence

39. T.H. Moray, J.E. Moray, The Sea of Energy, Cosray Research Institute, Salt Lake City, 1978.

40. T. Bearden, T. Herold, E. Mueller, "Gravity Field Generator Manufactured by John Bedini," Tesla Book Co., Greenville, Texas, 1985.

41. F. Sweet, T.E. Bearden, "Utilizing Scalar Electromagnetics to Tap Vacuum Energy," Proc. 26th IECEC, vol 4, 1991, pp 370-375; ... "Nothing is Something: The Theory and Operation of a Phase-Conjugate Vacuum Triode," privately published, 1988.

42. M.B. King, "Cohering the Zero-Point Energy," Proc. of the 1986 International Tesla Symposium, Colorado Springs, section 4, pp 13-32. Also in M.B. King, Tapping the Zero-Point Energy, Paraclete Publishing, Provo, Utah, 1989, pp 77-106.

Dual Vortex Forms: The Key to a Large Zero-Point Energy Coherence

Scalar Compression

September 2001

Introduction

A recurring theme in "free energy" inventions is to invoke a "scalar" activation of the zero-point energy by repeatedly triggering overlapping and opposing electric or magnetic fields in an abrupt, spiking manner [1]. The word "scalar" refers to the fact that the field vectors completely cancel leaving only a scalar potential. Popular approaches include pulsing caduceus wound coils [2] or rapidly accelerating opposing magnets past each other [3]. Support for the notion that anomalously excessive energy might arise from such activity comes from observations of the sun's corona [4]. Here solar flares can suddenly release up to 10^{25} joules when oppositely directed magnetic field lines in plasma filaments suddenly cross and cancel each other. The corona contains a turbulent, fractal distribution of solar flares. Could the anomalously high energy exhibited by the sun's corona actually be sourced from the zero-point energy? Could scalar compression be the key to easily creating a "free energy" device?

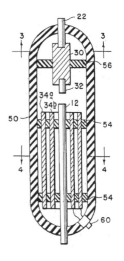

Sonoluminescence

Large zero-point energy activation might arise from an abrupt, symmetrical compression of matter or charged plasma. An example appears to be sonoluminescence, where the ultrasonic stimulation of water with dissolved inert gas exhibits a bluish glow, whose photons are emitted much too quickly to be from electron atomic transitions [5]. Eberlein [6] explains sonoluminescence arises because abrupt compression from collapsing bubbles converts zero-point energy directly into photons. Eberlein shows that any abrupt motion of matter activates the zero-point energy.

Symmetrical Charged Plasma Implosion

Perhaps even more powerfully, abrupt compression from an imploding symmetric shell of charged plasma might produce significant zero-point energy activation. Moreover, a tube designed to repeatedly create such activity could offer a means to electrically extract the energy: An electrode placed at the center of the implosion would experience a

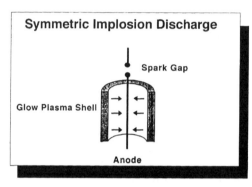

huge voltage spike. Excess energy might manifest as a pulse of vacuum polarized displacement current that would first surround the center electrode and then explosively propagate out along the connecting wires (which act like wave guides) to the external circuitry. The vacuum polarization pulse might manifest as "cold electricity" [7] if the transient propagates too fast for normal electron conduction to follow, and thus it would minimize ohmic heating. A tube that produced scalar compression from an abruptly imploding discharge, sourced from a symmetrical shell of charged glow plasma, might extract abundant zero-point energy.

Gray's Scalar Compression Tube

It appears that Gray [8] has invented a scalar compression tube (Figure 1). An outer double wall, cylindrical grid acts like a hollow cathode, which builds up negatively charged, glow plasma from a high voltage charging circuit. Hollow cathode tubes can provide pico second rise time switching transients [9]. The inner anode along the tube's axis is split to provide a spark gap. The spark initiates a photo-ionization avalanche breakdown in the gap between the cathode and anode, which triggers a cylindrical, symmetric, imploding shock wave [10] originating from the glow plasma. Gray channels the resulting high voltage spike to an "inductive load," which is surprising since such loads are known to exhibit high impedance to voltage spikes. However, a vacuum polarization pulse propagating across the outside of the coil might charge its windings by capacitive coupling.

Hyde's Pulse Current Multiplier

An alternative means to extract the energy from the high voltage spikes would be to use a pulse current multiplier (PCM) circuit [11] (Figure 2). The circuit was empirically invented by Hyde [12] to convert high voltage, low current, electrostatic spikes into high current, lower voltage pulses that standard rectification circuits could handle. Here each voltage spike propagates down the short distance, low impedance series path to charge the capacitors. The capacitors are drained in parallel through long leads that manifest blocking inductance to the original voltage spikes. The PCM circuit divides the voltage while multiplying the current. Hyde discovered that a rapid sequence of low power, high voltage spikes could be integrated to useful output power, and a net energy gain could manifest if the original spikes happen to capture zero-point energy.

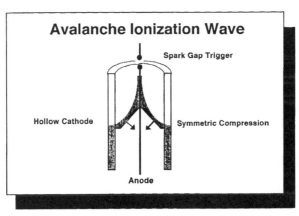

Diagram of an Avalanche Ionization Wave

The Gift

Both Gray and Hyde suffered greatly in their attempts to bring their inventions to the world. Would it not be ironic if simply combining Gray's scalar compression tube with Hyde's pulse current multiplier circuit would, in the end, gift mankind with unlimited energy?

Acknowledgement

The author wishes to thank Peter Lindemann for stimulating discussions.

References

1. M.B. King (1986), "Cohering the Zero-Point Energy," *Proc. of the 1986 International Tesla Symposium,* Colorado Springs, section 4, pp 13-32.

2. M.B. King (1988), "Demonstrating a Zero-Point Energy Coherence," *Proc. of the 1988 International Tesla Symposium, Colorado Springs,* section 4, pp 1-13.

3. M.B. King (1998), "Vortex Filaments, Torsion Fields and the Zero-Point Energy," *J. New Energy,* vol 3, no 2/3, pp 106-116.

4. B.N. Dwivedi, K.J.H. Phillips (June 2001), "The Paradox of the Sun's Hot Corona," *Sci. Amr.* vol 84, no 6, pp 40-47.

5. B.P. Barber, S.J. Putterman (1991), "Observation of synchronous picosecond sonoluminescence," *Nature,* vol 353, pp 318-320. Also (1992), "Light Scattering Measurements of the Repetitive Supersonic Implosion of a Sonoluminescing Bubble," *Phys. Rev. Lett.* vol 69, pp 3839-42.

6. C. Eberlein (1996), "Sonoluminescence as Quantum Vacuum Radiation," *Phys. Rev. Lett.* vol 76, pp 3842-45. Also (1996), "Theory of quantum radiation observed as sonoluminescence," *Phys. Rev. A,* vol 53, pp 2772-87.

7. P.A. Lindemann (2001), The Free Energy Secrets of Cold Electricity, Clear Tech Inc., Metaline Falls, WA. Website: www.free-energy.cc

8. E.V. Gray (1987), "Efficient Electrical Conversion Switching Tube Suitable for Inductive Loads," U.S. Patent 4,661,747. Also (1986), "Efficient Power Supply Suitable for Inductive Loads," U.S. Patent 4,595,975.

9. M.A. Gundersen, G. Schaefer (1990), Physics and Applications of Pseudosparks, Plenum Press, NY.

10. A.N. Lagarkov, I.M. Rutkevich (1994), Ionization Waves in Electrical Breakdown of Gases, Springer-Verlag, NY, p 100.

11. M.B. King (1996), "The Super Tube," *Proc. Int. Sym. on New Energy,* pp 209-221. Also (1997), *Infinite Energy,* vol 2, no 8, pp 23-28.

12. W.W. Hyde (1990), "Electrostatic Energy Field Power Generating System," U.S. Patent No. 4,897,592.

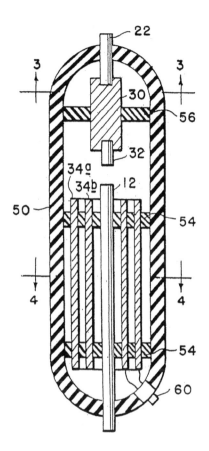

Figure 1 Gray's scalar compression tube from U.S. Patent 4,661,747. The hollow cathode'34 (*a & b*) contains negatively charged glow plasma. A sharp, unipolar spark from anode 12 to 32 triggers avalanche breakdown in the cylindrical gap between cathode 34 and anode 12 resulting in a symmetrical, imploding ionization shock wave originating from the glow plasma. The ionization wave front propagates downward in the cylindrical gap to launch a negative, vacuum polarization transient out the bottom of the tube guided on electrode 12.

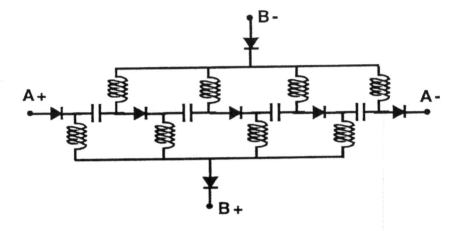

Figure 2 Pulse Current Multiplier (PCM) circuit converts unipolar voltage spikes (input on the *A* terminals) to current pulses (output on the *B* terminals). Four stages are shown, but more can be used. The current multiplication (and voltage division) is proportional to the number of stages. For sharp voltage spikes, long leads can be used in place of the inductors as long as the series path between the input terminals is a short distance.

United States Patent [19]

Gray, Sr.

[11] **Patent Number:** **4,661,747**

[45] **Date of Patent:** **Apr. 28, 1987**

[54] **EFFICIENT ELECTRICAL CONVERSION SWITCHING TUBE SUITABLE FOR INDUCTIVE LOADS**

[76] Inventor: **Edwin V. Gray, Sr.**, P.O. Box 362, Council, Id. 83612

[21] Appl. No.: **791,508**

[22] Filed: **Oct. 25, 1985**

Related U.S. Application Data

[62] Division of Ser. No. 662,339, Oct. 18, 1984, Pat. No. 4,595,975.

[30] **Foreign Application Priority Data**

Dec. 16, 1983 [GR] Greece 124388

[51] Int. Cl.⁴ **H01J 11/04; H01J 13/48; H05B 37/00; H05B 39/00**

[52] U.S. Cl. **315/330;** 313/601; 313/602; 313/604; 315/261; 315/335

[58] **Field of Search** 315/57, 58, 60, 36, 315/334, 335, 330, 336, 261; 313/595, 601, 602, 603

[56] **References Cited**

U.S. PATENT DOCUMENTS

3,443,142	5/1969	Koppl et al.	315/58
3,663,855	5/1972	Boettcher	315/330
3,798,461	3/1974	Edson	315/36
3,939,379	2/1976	Sullivan et al.	315/330
4,198,590	4/1980	Harris	315/335
4,370,597	1/1983	Weiner et al.	315/58

FOREIGN PATENT DOCUMENTS

0540361 12/1976 U.S.S.R. 315/335

Primary Examiner—Saxfield Chatmon

[57] **ABSTRACT**

Disclosed is an electrical driving and recovery system for a high frequency environment. The recovery system can be applied to drive present day direct-current or alternating-current loads for better efficiency. It has a low-voltage source coupled to a vibrator, a transformer and a bridge-type rectifier to provide a high voltage pulsating signal to a first capacitor. Where a high-voltage source is otherwise available, it may be coupled directly to a bridge-type rectifier, causing a pulsating signal to the first capacitor. The first capacitor in turn is coupled to a high voltage anode of an electrical conversion switching element tube. The switching element tube also includes a low voltage anode which is connected to a voltage source by a commutator and a switching element tube. Mounted around the high voltage anode is a charge receiving plate which is coupled to an inductive load to transmit a high voltage discharge from the switching element tube to the load. Also coupled to the load is a second capacitor for storing the back EMF created by the collapsing electrical field of the load when the current to the load is blocked. The second capacitor is coupled to the voltage source. When adapted to present day direct-current or alternating-current devices the load could be a battery or capacitor to enhance the productivity of electrical energy.

4 Claims, 5 Drawing Figures

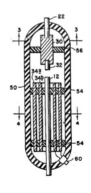

FIG. 1

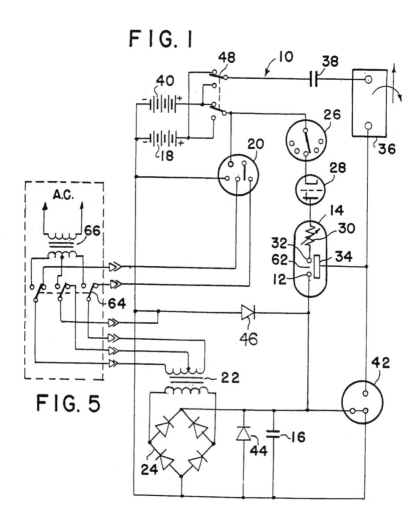

FIG. 5

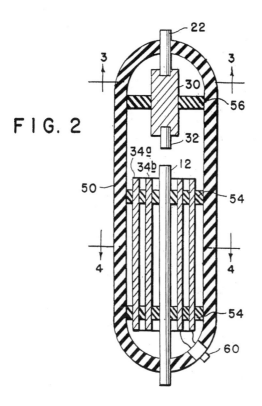

FIG. 2

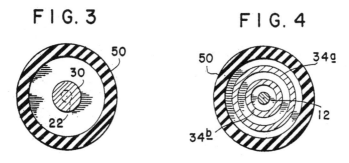

FIG. 3

FIG. 4

1

EFFICIENT ELECTRICAL CONVERSION SWITCHING TUBE SUITABLE FOR INDUCTIVE LOADS

This is a division of application Ser. No. 662,339, filed Oct. 18, 1984, now U.S. Pat. No. 4,595,975.

BACKGROUND OF THE INVENTION

1. Field of the Invention

The present invention relates to an electrical driving system and a conversion element, and more particularly, to a system for driving an inductive load in a greatly improved and efficient manner.

2. Description of the Prior Act

In the opinion of the inventor, there is no known device which provides the conversion of energy from a direct-current electric source or an alternating-current electric source to a mechanical force based on the principle of this invention. EXAMPLE: A portable energy source, (1) such as a battery, (2) such as alternating-current, (3) such as the combination of battery and alternating-current, may be used with highly improved efficiency to operate a mechanical device, whose output is a linear or rotary force, with an attendant increase in the useful productive period between external applications of energy restoration to the energy source.

SUMMARY OF THE INVENTION

The present invention provides a more efficient driving system comprising a source of electrical voltage; a vibrator connected to the low-voltage source for forming a pulsating signal; a transformer connected to the vibrator for receiving the pulsating signal; a high-voltage source, where available, connected to a bridge-type rectifier; or the bridge-type rectifier connected to the high voltage pulse output of the transformer; a capacitor for receiving the voltage pulse output; a conversion element having first and second anodes, electrically conductive means for receiving a charge positioned about the second anode and an output terminal connected to the charge receiving means, the second anode being connected to the capacitor; a commutator connected to the source of electrical voltage and to the first anode; and an inductive load connected to the output terminal whereby a high energy discharge between the first and second anodes is transferred to the charge receiving means and then to the inductive load.

As a sub-combination, the present invention also includes a conversion element comprising a housing; a first low voltage anode mounted to the housing, the first anode adapted to be connected to a voltage source; a second high voltage anode mounted to the housing, the second anode adapted to be connected to a voltage source; electrically conductive means positioned about the second anode and spaced therefrom for receiving a charge, the charge receiving means being mounted to the housing; and an output terminal communicating with the charge receiving means, said terminal adapted to be connected to an inductive load.

The invention also includes a method for providing power to an inductive load comprising the steps of providing a voltage source, pulsating a signal from said source; increasing the voltage of said signal; rectifying said signal; storing and increasing the signal; conducting said signal to a high voltage anode; providing a low voltage to a second anode to form a high energy discharge; electrostatically coupling the discharge to a charge receiving element; conducting the discharge to an inductive load; coupling a second capacitor to the load; and coupling the second capacitor to the source.

It is an aim of the present invention to provide a system for driving an inductive load which system is substantially more efficient than any now existing.

Another object of the present invention is to provide a system for driving an inductive load which is reliable, is inexpensive and simply constructed.

The foregoing objects of the present invention together with various other objects, advantages, features and results thereof which will be evident to those skilled in the art in light of this disclosure may be achieved with the exemplary embodiment of the invention described in detail hereinafter and illustrated in the accompanying drawings.

BRIEF DESCRIPTION OF THE DRAWINGS

FIG. 1 is a schematic circuit diagram of the electrical driving system.

FIG. 2 is an elevational sectional view of the electrical conversion element.

FIG. 3 is a plan sectional view taken along line 3—3 of FIG. 2.

FIG. 4 is a plan sectional view taken along line 4—4 of FIG. 2.

FIG. 5 is a schematic circuit diagram of the alternating-current input circuit.

DESCRIPTION OF THE PREFERRED EMBODIMENT

While the present invention is susceptible of various modifications and alternative constructions, an embodiment is shown in the drawings and will herein be described in detail. It should be understood however that it is not the intention to limit the invention to the particular form disclosed; but, on the contrary, the invention is to cover all modifications, equivalents and alternative constructions falling within the spirit and scope of the invention as expressed in the appended claims.

There is disclosed herein an electrical driving system which, on theory, will convert low voltage electric energy from a source such as an electric storage battery to a high potential, high current energy pulse that is capable of developing a working force at the inductive output of the device that is more efficient than that which is capable of being developed directly from the energy source. The improvement in efficiency is further enhanced by the capability of the device to return that portion of the initial energy developed, and not used by the inductive load in the production of mechanical energy, to the same or second energy reservoir or source for use elsewhere, or for storage.

This system accomplishes the results stated above by harnessing the "electrostatic" or "impulse" energy created by a high-intensity spark generated within a specially constructed electrical conversion switching element tube. This element utilizes a low-voltage anode, a high-voltage anode, and one or more "electrostatic" or charge receiving grids. These grids are of a physical size, and appropriately positioned, as to be compatible with the size of the tube, and therefore, directly related to the amount of energy to be anticipated when the device is operating.

The low-voltage anode may incorporate a resistive device to aid in controlling the amount of current drawn from the energy source. This low-voltage anode is connected to the energy source through a mechanical commutator or a solid-state pulser that controls the timing and duration of the energy spark within the element. The high-voltage anode is connected to a high-voltage potential developed by the associated circuits. An energy discharge occurs within the element when the external control circuits permit. This short duration, high-voltage, high-current energy pulse is captured by the "electrostatic" grids within the tube, stored momentarily, then transferred to the inductive output load.

The increase in efficiency anticipated in converting the electrical energy to mechanical energy within the inductive load is attributed to the utilization of the most optimum timing in introducing the electrical energy to the load device, for the optimum period of time.

Further enhancement of energy conservation is accomplished by capturing a significant portion of the energy generated by the inductive load when the useful energy field is collapsing. This energy is normally dissipated in load losses that are contrary to the desired energy utilization, and have heretofore been accepted because no suitable means had been developed to harness this energy and restore it to a suitable energy storage device.

The present invention is concerned with two concepts or characteristics. The first of these characteristics is observed with the introduction of an energizing current through the inductor. The inductor creates a contrary force (counter-electromotive force or CEMF) that opposes the energy introduced into the inductor. This CEMF increases throughout the time the introduced energy is increasing.

In normal applications of an alternating-current to an inductive load for mechanical applications, the useful work of the inductor is accomplished prior to terminating the application of energy. The excess energy applied is thereby wasted.

Previous attempts to provide energy inputs to an inductor of time durations limited to that period when the optimum transfer of inductive energy to mechanical energy is occuring, have been limited by the ability of any such device to handle the high current required to optimize the energy transfer.

The second characteristic is observed when the energizing current is removed from the inductor. As the current is decreased, the inductor generates an EMF that opposes the removal of current or, in other words, produces an energy source at the output of the inductor that simulates the original energy source, reduced by the actual energy removed from the circuit by the mechanical load. This "regenerated", or excess, energy has previously been lost due to a failure to provide a storage capability for this energy.

In this invention, a high-voltage, high-current, short duration energy pulse is applied to the inductive load by the conversion element. This element makes possible the use of certain of that energy impressed within an arc across a spark-gap, without the resultant deterioration of circuit elements normally associated with high energy electrical arcs.

This invention also provides for capture of a certain portion of the energy induced by the high inductive kick produced by the abrupt withdrawal of the introduced current. This abrupt withdrawal of current is attendant upon the termination of the stimulating arc. The voltage spike so created is imposed upon a capacitor that couples the attendant current to a secondary energy storage device.

A novel, but not essential, circuit arrangement provides for switching the energy source and the energy storage device. This switching may be so arranged as to actuate automatically at predetermined times. The switching may be at specified periods determined by experimentation with a particular device, or may be actuated by some control device that measures the relative energy content of the two energy reservoirs.

Referring now to FIG. 1, the system 10 will be described in additional detail. The potential for the high-voltage anode 12 of the conversion element 14 is developed across the capacitor 16. This voltage is produced by drawing a low current from a battery source 18 through the vibrator 20. The effect of the vibrator is to create a pulsating input to the transformer 22. The turns ratio of the transformer is chosen to optimize the voltage applied to a bridge-type rectifier 24. The output of the rectifier is then a series of high-voltage pulses of modest current. When the available source is already of the high voltage AC type, it may be coupled directly to the bridge-type rectifier.

By repetitious application of these output pulses from the bridge-type recrifier to the capacitor 16, a high-voltage high-level charge is built up on the capacitor.

Control of the conversion switching element tube is maintained by a commutator 26. A series of contacts mounted radially about a shaft, or a solid-state switching device sensitive to time or other variable may be used for this control element. A switching element tube type one-way energy path 28 is introduced between the commutator device and the conversion switching element tube to prevent high energy arcing at the commutator current path. When the switching element tube is closed, current from the voltage source 18 is routed through a resistive element 30 and a low voltage anode 32. This causes a high energy discharge between the anodes within the conversion switching element tube 14.

The energy content of the high energy pulse is electrostatically coupled to the conversion grids 34 of the conversion element. This electrostatic charge is applied through an output terminal 60 (FIG. 2) across the load inductance 36, inducing a strong electromagnetic field about the inductive load. The intensity of this electromagnetic field is determined by the high electromotive potential developed upon the electrostatic grids and the very short time duration required to develop the energy pulse.

If the inductive load is coupled magnetically to a mechanical load, a strong initial torque is developed that may be efficiently utilized to produce physical work.

Upon cessation of the energy pulse (arc) within the conversion switching element tube the inductive load is decoupled, allowing the electromagnetic field about the inductive load to collapse. The collapse of this energy field induces within the inductive load a counter EMF. This counter EMF creates a high positive potential across a second capacitor 38 which, in turn, is induced into the second energy storage device or battery 40 as a charging current. The amount of charging current available to the battery 40 is dependent upon the initial conditions within the circuit at the time of discharge within the conversion switching element tube and the amount of mechanical energy consumed by the work load.

A spark-gap protection device 42 is included in the circuit to protect the inductive load and the rectifier

5

elements from unduly large discharge currents. Should the potentials within the circuit exceed predetermined values, fixed by the mechanical size and spacing of the elements within the protective device, the excess energy is dissipated (bypassed) by the protective device to the circuit common (electrical ground).

Diodes 44 and 46 bypass the excess overshoot generated when the "Energy Conversion Switching Element Tube" is triggered. A switching element 48 allows either energy storage source to be used as the primary energy source, while the other battery is used as the energy retrieval unit. The switch facilitates interchanging the source and the retrieval unit at optimum intervals to be determined by the utilization of the conversion switching element tube. This switching may be accomplished manually or automatically, as determined by the choice of switching element from among a large variety readily available for the purpose.

FIGS. 2, 3, and 4 show the mechanical structure of the conversion switching element tube 14. An outer housing 50 may be of any insulative material such as glass. The anodes 12 and 32 and grids 34a and 34b are firmly secured by nonconductive spacer material 54, and 56. The resistive element 30 may be introduced into the low-voltage anode path to control the peak currents through the conversion switching element tube. The resistive element may be of a piece, or it may be built of one or more resistive elements to achieve the desired result.

The anode material may be identical for each anode, or may be of differing materials for each anode, as dictated by the most efficient utilization of the device, as determined by appropriate research at the time of production for the intended use.

The shape and spacing of the electrostatic grids is also susceptible to variation with application (voltage, current, and energy requirements).

It is the contention of the inventor that by judicious mating of the elements of the conversion switching element tube, and the proper selection of the components of the circuit elements of the system, the desired theoretical results may be achieved. It is the inventor's contention that this mating and selection process is well within the capabilities of intensive research and development technique.

Let it be stated here that substituting a source of electric alternating-current subject to the required current and/or voltage shaping and/or timing, either prior to being considered a primary energy source, or thereafter, should not be construed to change the described utilization or application of primary energy in any way. Such energy conversion is readily achieved by any of a multitude of well established principles. The preferred embodiment of this invention merely assumes optimum utilization and optimum benefit from this invention when used with portable energy devices similar in principle to the wet-cell or dry-cell battery.

This invention proposes to utilize the energy contained in an internally generated high-voltage electric spike (energy pulse) to electrically energize an inductive load; this inductive load being then capable of converting the energy so supplied into a useful electrical or mechanical output

In operation the high-voltage, short-duration electric spike is generated by discharging the capacitor 16 across the spark-gap in the conversion switching element tube. The necessary high-voltage potential is

6

stored on the capacitor in incremental, additive steps from the bridge-type rectifier 24.

When the energy source is a direct-current electric energy storage device, such as the battery 12, the input to the bridge rectifier is provided by the voltage step-up transformer 22, that is in turn energized from the vibrator 20, or solid-state chopper, or similar device to properly drive the transformer and rectifier circuits.

When the energy source is an alternating-current, switches 64 disconnect transformer 22 and the input to the bridge-type rectifier 24 is provided by the voltage step-up transformer 66, that is in turn energized from the vibrator 20, or solid-state chopper, or similar device to properly drive the transformer and rectifier circuits.

The repetitious output of the bridge rectifier incrementally increases the capacitor charge toward its maximum. This charge is electrically connected directly to the high-voltage anode 12 of the conversion switching element tube.

When the low-voltage anode 32 is connected to a source of current, an arc is created in the spark-gap designated 62 of the conversion switching element tube equivalent to the potential stored on the high-voltage anode, and the current available from the low-voltage anode. Because the duration of the arc is very short, the instantaneous voltage, and instantaneous current may both be very high. The instantaneous peak apparent power is therefore, also very high. Within the conversion switching element tube, this energy is absorbed by the grids 34a and 34b mounted circumferentially about the interior of the tube.

Control of the energy spike within the conversion switching element tube is accomplished by a mechanical, or solid-state commutator, that closes the circuit path from the low-voltage anode to the current source at that moment when the delivery of energy to the output load is most auspicious. Any number of standard high-accuracy, variable setting devices are available for this purpose. When control of the repetitive rate of system's output is required, it is accomplished by controlling the time of connection at the low-voltage anode.

Thus there can be provided an electrical driving system having a low-voltage source coupled to a vibrator, a transformer and a bridge-type rectifier to provide a high voltage pulsating signal to a first capacitor. Where a high-voltage source is otherwise available, it may be coupled direct to a bridge-type rectifier, causing a pulsating signal to a first capacitor. The capacitor in turn is coupled to a high-voltage anode of an electrical conversion switching element tube. The element also includes a low-voltage anode which in turn is connected to a voltage source by a commutator, a switching element tube, and a variable resistor. Mounted around the high-voltage anode is a charge receiving plate which in turn is coupled to an inductive load to transmit a high-voltage discharge from the element to the load. Also coupled to the load is a second capacitor for storing the back EMF created by the collapsing electrical field of the load when the current to the load is blocked. The second capacitor in turn is coupled to the voltage source.

What is claimed is:

1. An electrical conversion switching element tube comprising:

a closed insulative housing (50);

a first low-voltage anode (32) mounted internally to said housing and extending internally to an electri-

7

cal discharge area (62), said first anode adapted to be connected to a voltage source external to the housing;

a second high-voltage anode (12) mounted internally to said housing and extending internally to said electrical discharge area (62), said second anode also being adapted to be connected to a voltage source external to the housing;

electrically conductive means (34b) positioned internally within said housing and extending circumferentially about said second anode while being directly exposed thereto but not conductively connected thereto but, rather, spaced therefrom for receiving an electrostatic charge from the second anode when a discharge current is triggered across said discharge area between said first and second anodes, said charge receiving electrically conduc-

8

tive means also being internally mounted to said housing; and

an output terminal (60) communicating with said charge receiving electrically conductive means, said terminal adapted to be connected to an inductive load externally of said housing.

2. An electrical conversion switching element tube as claimed in claim 1, including a resistive element (30) in series with said first anode.

3. An electrical conversion switching element tube as claimed in claim 1 wherein:

said charge receiving electrically conductive means is tubularly shaped.

4. An electrical conversion switching element tube as claimed in claim 3, including

a second tubularly shaped charge receiving electrically conductive means (34a) positioned circumferentially about said first mentioned charge receiving electrically conductive means.

* * * * *

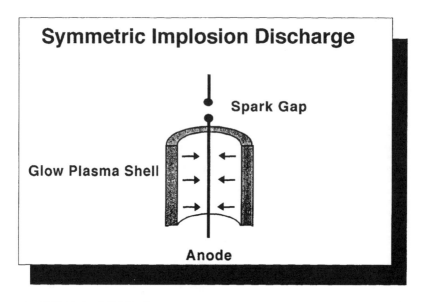

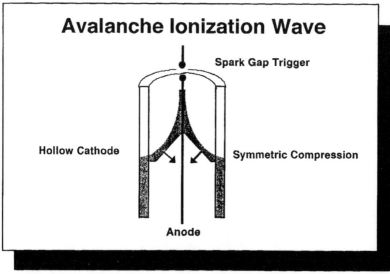

Figures A (top) and B (Bottom) These figures support the explanation of Gray's tube in Figure 1.

Afterword

The critical question is why hasn't the discovery of zero-point energy devices already been delivered to the world? Most career scientists, engineers and professors refuse to research the field since it violates the reigning paradigm: They typically believe the vacuum energy does not exist (or they believe it could never be a source of appreciable energy). A paradigm shift requires an experiment. Where is the evidence? Most claims by inventors typically involve measurement only. Since energy measurements are notoriously difficult (especially involving pulsed systems), the claims are generally ignored. After all if a device can output excessive power three or more times the input, why not close the loop and make a free-running system? The engineering to do so is usually costly. Many projects are typically under funded, and often succumb to questionable or even fraudulent business practices. Litigation is easy since experts from academia can authoritatively state any over-unity claim must be a fraud. Those few inventors that achieved enough excessive power to make their device free running and therefore convincing to the world, as well as had proper business practices enabling them to proceed to a manufacturing level, invariably ran into a wall: They were suppressed.

Who is behind the suppression? Is today's entrenched industry just trying to maintain the status quo economy? Is the government concerned that a new energy discovery could be abused? While the former is just motivated by profits, the latter has a valid point. If Wheeler's description is right and we have access to 10^{94} g/cm^3 of energy, the situation is dramatically double edged. We can use the discovery to make utopia by meeting all of mankind's physical needs, expanding to a glorious future. (The alternative is too grim.) Is mankind ready for the discovery?

A glance at the news showing man's abuse to man makes it apparent that we are not ready. Yet the continued use of fossil fuels will likely cause our ecological destruction within 50 years. How do we solve the problem responsibly? One possible solution is that the discovery be kept secret and controlled by a central authority like atomic power. Until we grow up spiritually we would not have complete freedom for energy use. Another potential solution arises from the consciousness movement: Mankind awakens to the awareness that we are connected, and we are part of a greater spiritual being. Ideally the awakening would make us a loving, empathetic species that might even be able to communicate by telepathy. Could such an optimistic outcome be possible?

I have participated in some consciousness courses and have met many inspired people who feel they were born at this time to help elevate mankind into a new spiritual state. I have personally experienced so many coincidental, synchronistic events that I feel like I'm living in a world as described in Redfield's book, *The Celestine Prophecy.* For me the most significant coincidence was my name. In 1976, after I was convinced that the zero-point energy could potentially become a new energy source, I was given the book, *The Sea of Energy in which the Earth Floats* by T. Henry Moray. Upon seeing the name, I was stunned. Moray's free energy invention was perhaps the most widely witnessed device in the history of the field. At that point I knew my life's purpose was to explain the energy ideas so well that we would rediscover the fundamental source of energy.

Since that time, I have experienced so many other synchronicities that I know we are being spiritually guided. People all over the globe are now stepping up in both the consciousness movement and the world's religions to help elevate our awareness and spirituality. I believe we are going to make it.

It just might be that the discovery of the vacuum energy as a limitless energy source is to be synchronized with a spiritual renaissance of all of humanity. If so, the quest for zero-point energy broadens into a consciousness transformation at the planetary level. Many feel their life's purpose is to help with this transition.

You know who you are. If you feel this way, welcome to the quest!

About the Author

Moray B. King has a B.S. in Electrical Engineering (1972), M.S. in Systems Engineering (1976), and has completed the Ph.D. courses in Systems Engineering (on thesis leave) all from the University of Pennsylvania. Since 1974 he has been researching the standard physics literature on zero-point energy and has studied the experiments of many scientists and inventors who have reported anomalous energy gains. He has given numerous presentations at energy conferences, written many technical papers, and authored the book, *Tapping the Zero-Point Energy*. In his career he has been employed as a system engineer, software engineer, research consultant, and is currently the senior scientist of Paraclete Corporation.

ALSO BY MORAY B. KING

TAPPING THE ZERO POINT ENERGY
Free Energy & Anti-Gravity in Today's Physics

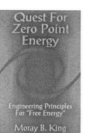

QUEST FOR ZERO-POINT ENERGY
Engineering Principles for "Free Energy"
by Moray B. King
King expands, with diagrams, on how free energy and anti-gravity are possible. The theories of zero point energy maintain there are tremendous fluctuations of electrical field energy embedded within the fabric of space. King explains the following topics: Tapping the Zero-Point Energy as an Energy Source; Fundamentals of a Zero-Point Energy Technology; Vacuum Energy Vortices; The Super Tube; Charge Clusters: The Basis of Zero-Point Energy Inventions; Vortex Filaments, Torsion Fields and the Zero-Point Energy; Transforming the Planet with a Zero-Point Energy Experiment; Dual Vortex Forms: The Key to a Large Zero-Point Energy Coherence. Packed with diagrams, patents and photos. With power shortages now a daily reality in many parts of the world, this book offers a fresh approach very rarely mentioned in the mainstream media.
224 PAGES. 6x9 PAPERBACK. ILLUSTRATED. $14.95. CODE: QZPE

TAPPING THE ZERO POINT ENERGY
Free Energy & Anti-Gravity in Today's Physics
by Moray B. King
King explains how free energy and anti-gravity are possible. The theories of the zero point energy maintain there are tremendous fluctuations of electrical field energy imbedded within the fabric of space. This book tells how, in the 1930s, inventor T. Henry Moray could produce a fifty kilowatt "free energy" machine; how an electrified plasma vortex creates anti-gravity; how the Pons/Fleischmann "cold fusion" experiment could produce tremendous heat without fusion; and how certain experiments might produce a gravitational anomaly.
180 PAGES. 5x8 PAPERBACK. ILLUSTRATED. $12.95. CODE: TAP

THE FREE-ENERGY DEVICE HANDBOOK
A Compilation of Patents and Reports
by David Hatcher Childress

A large-format compilation of various patents, papers, descriptions and diagrams concerning free-energy devices and systems. The Free-Energy Device Handbook is a visual tool for experimenters and researchers into magnetic motors and other "over-unity" devices. With chapters on the Adams Motor, the Hans Coler Generator, cold fusion, superconductors, "N" machines, space-energy generators, Nikola Tesla, T. Townsend Brown, and the latest in free-energy devices. Packed with photos, technical diagrams, patents and fascinating information, this book belongs on every science shelf. With energy and profit being a major political reason for fighting various wars, free-energy devices, if ever allowed to be mass distributed to consumers, could change the world! Get your copy now before the Department of Energy bans this book!
292 PAGES. 8x10 PAPERBACK. ILLUSTRATED. BIBLIOGRAPHY. $16.95. CODE: FEH

ETHER TECHNOLOGY
A Rational Approach to Gravity Control
by Rho Sigma

This classic book on anti-gravity and free energy is back in print and back in stock. Written by a well-known American scientist under the pseudonym of "Rho Sigma," this book delves into international efforts at gravity control and discoid craft propulsion. Before the Quantum Field, there was "Ether." This small, but informative book has chapters on John Searle and "Searle discs;" T. Townsend Brown and his work on anti-gravity and ether-vortex turbines. Includes a forward by former NASA astronaut Edgar Mitchell.
108 PAGES. 6x9 PAPERBACK. ILLUSTRATED. $12.95. CODE: ETT

HARNESSING THE WHEELWORK OF NATURE
Tesla's Science of Energy
by Thomas Valone, Ph.D., P.E.

A compilation of essays, papers and technical briefings on the emerging Tesla Technology and Zero Point Energy engineering that will soon change the entire way we live. Chapters include: Tesla: Scientific Superman who Launched the Westinghouse Industrial Firm by John Shatlan; Nikola Tesla—Electricity's Hidden Genius, excerpt from The Search for Free Energy; Tesla's History at Niagara Falls; Non-Hertzian Waves: True Meaning of the Wireless Transmission of Power by Toby Grotz; On the Transmission of Electricity Without Wires by Nikola Tesla; Tesla's Magnifying Transmitter by Andrija Puharich; Tesla's Self-Sustaining Electrical Generator and the Ether by Oliver Nichelson; Self-Sustaining Non-Hertzian Longitudinal Waves by Dr. Robert Bass; On the Transmission of Electricity in Free Space; Scalar Electromagnetic Waves; Disclosures Concerning Tesla's Operation of an ELF Oscillator; A Study of Tesla's Advanced Concepts & Glossary of Tesla Technology Terms; Electric Weather Forces: Tesla's Vision by Charles Yost; The New Art of Projecting Concentrated Non-Dispersive Energy Through Natural Media; The Homopolar Generator: Tesla's Contribution by Thomas Valone; Tesla's Ionizer and Ozonator: Implications for Indoor Air Pollution by Thomas Valone; How Cosmic Forces Shape Our Destiny by Nikola Tesla; Tesla's Death Ray plus Selected Tesla Patents; more.
288 PAGES. 6x9 PAPERBACK. ILLUSTRATED. $16.95. CODE: HWWN

THE ANTI-GRAVITY HANDBOOK
edited by David Hatcher Childress, with Nikola Tesla, T.B. Paulicki, Bruce Cathie, Albert Einstein and others

The new expanded compilation of material on Anti-Gravity, Free Energy, Flying Saucer Propulsion, UFOs, Suppressed Technology, NASA Cover-ups and more. Highly illustrated with patents, technical illustrations and photos. This revised and expanded edition has more material, including photos of Area 51, Nevada, the government's secret testing facility. This classic on weird science is back in a 90s format!
• **How to build a flying saucer.**
•**Arthur C. Clarke on Anti-Gravity.**
• **Crystals and their role in levitation.**
• **Secret government research and development.**
• **Nikola Tesla on how anti-gravity airships could draw power from the atmosphere.**
• **Bruce Cathie's Anti-Gravity Equation.**
• **NASA, the Moon and Anti-Gravity.**
230 PAGES. 7x10 PAPERBACK. ILLUSTRATED. $14.95. CODE: **AGH**

ANTI–GRAVITY & THE WORLD GRID

Is the earth surrounded by an intricate electromagnetic grid network offering free energy? This compilation of material on ley lines and world power points contains chapters on the geography, mathematics, and light harmonics of the earth grid. Learn the purpose of ley lines and ancient megalithic structures located on the grid. Discover how the grid made the Philadelphia Experiment possible. Explore the Coral Castle and many other mysteries, including acoustic levitation, Tesla Shields and scalar wave weaponry. Browse through the section on anti-gravity patents, and research resources.
274 PAGES. 7x10 PAPERBACK. ILLUSTRATED. $14.95. CODE: **AGW**

ANTI–GRAVITY & THE UNIFIED FIELD
edited by David Hatcher Childress

Is Einstein's Unified Field Theory the answer to all of our energy problems? Explored in this compilation of material is how gravity, electricity and magnetism manifest from a unified field around us. Why artificial gravity is possible; secrets of UFO propulsion; free energy; Nikola Tesla and anti-gravity airships of the 20s and 30s; flying saucers as superconducting whirls of plasma; anti-mass generators; vortex propulsion; suppressed technology; government cover-ups; gravitational pulse drive; spacecraft & more.
240 PAGES. 7x10 PAPERBACK. ILLUSTRATED. $14.95. CODE: **AGU**

THE TESLA PAPERS
Nikola Tesla on Free Energy & Wireless Transmission of Power
by Nikola Tesla, edited by David Hatcher Childress

David Hatcher Childress takes us into the incredible world of Nikola Tesla and his amazing inventions. Tesla's rare article "The Problem of Increasing Human Energy with Special Reference to the Harnessing of the Sun's Energy" is included. This lengthy article was originally published in the June 1900 issue of *The Century Illustrated Monthly Magazine* and it was the outline for Tesla's master blueprint for the world. Tesla's fantastic vision of the future, including wireless power, anti-gravity, free energy and highly advanced solar power. Also included are some of the papers, patents and material collected on Tesla at the Colorado Springs Tesla Symposiums, including papers on: •The Secret History of Wireless Transmission •Tesla and the Magnifying Transmitter •Design and Construction of a Half-Wave Tesla Coil •Electrostatics: A Key to Free Energy •Progress in Zero-Point Energy Research •Electromagnetic Energy from Antennas to Atoms •Tesla's Particle Beam Technology •Fundamental Excitatory Modes of the Earth-Ionosphere Cavity
325 PAGES. 8x10 PAPERBACK. ILLUSTRATED. $16.95. CODE: **TTP**

THE FANTASTIC INVENTIONS OF NIKOLA TESLA
by Nikola Tesla with additional material by David Hatcher Childress

This book is a readable compendium of patents, diagrams, photos and explanations of the many incredible inventions of the originator of the modern era of electrification. In Tesla's own words are such topics as wireless transmission of power, death rays, and radio-controlled airships. In addition, rare material on German bases in Antarctica and South America, and a secret city built at a remote jungle site in South America by one of Tesla's students, Guglielmo Marconi. Marconi's secret group claims to have built flying saucers in the 1940s and to have gone to Mars in the early 1950s! Incredible photos of these Tesla craft are included. The Ancient Atlantean system of broadcasting energy through a grid system of obelisks and pyramids is discussed, and a fascinating concept comes out of one chapter: that Egyptian engineers had to wear protective metal head-shields while in these power plants, hence the Egyptian Pharoah's head covering as well as the Face on Mars! •His plan to transmit free electricity into the atmosphere. •How electrical devices would work using only small antennas. •Why unlimited power could be utilized anywhere on earth. •How radio and radar technology can be used as death-ray weapons in Star Wars.

342 PAGES. 6x9 PAPERBACK. ILLUSTRATED. $16.95. CODE: **FINT**

LOST SCIENCE
by Gerry Vassilatos
Rediscover the legendary names of suppressed scientific revolution—remarkable lives, astounding discoveries, and incredible inventions which would have produced a world of wonder. How did the aura research of Baron Karl von Reichenbach prove the vitalistic theory and frighten the greatest minds of Germany? How did the physiophone and wireless of Antonio Meucci predate both Bell and Marconi by decades? How does the earth battery technology of Nathan Stubblefield portend an unsuspected energy revolution? How did the geoaetheric engines of Nikola Tesla threaten the establishment of a fuel-dependent America? The microscopes and virus-destroying ray machines of Dr. Royal Rife provided the solution for every world-threatening disease. Why did the FDA and AMA together condemn this great man to Federal Prison? The static crashes on telephone lines enabled Dr. T. Henry Moray to discover the reality of radiant space energy. Was the mysterious "Swedish stone," the powerful mineral which Dr. Moray discovered, the very first historical instance in which stellar power was recognized and secured on earth? Why did the Air Force initially fund the gravitational warp research and warp-cloaking devices of T. Townsend Brown and then reject it? When the controlled fusion devices of Philo Farnsworth achieved the "break-even" point in 1967 the FUSOR project was abruptly cancelled by ITT.
304 PAGES. 6X9 PAPERBACK. ILLUSTRATED. BIBLIOGRAPHY. $16.95. CODE: LOS

SECRETS OF COLD WAR TECHNOLOGY
Project HAARP and Beyond
by Gerry Vassilatos
Vassilatos reveals that "Death Ray" technology has been secretly researched and developed since the turn of the century. Included are chapters on such inventors and their devices as H.C. Vion, the developer of auroral energy receivers; Dr. Selim Lemstrom's pre-Tesla experiments; the early beam weapons of Grindell-Mathews, Ulivi, Turpain and others; John Hettenger and his early beam power systems. Learn about Project Argus, Project Teak and Project Orange; EMP experiments in the 60s; why the Air Force directed the construction of a huge Ionospheric "backscatter" telemetry system across the Pacific just after WWII; why Raytheon has collected every patent relevant to HAARP over the past few years; more.
250 PAGES. 6X9 PAPERBACK. ILLUSTRATED. $15.95. CODE: SCWT

THE ENERGY GRID
Harmonic 695, The Pulse of the Universe
by Captain Bruce Cathie.
This is the breakthrough book that explores the incredible potential of the Energy Grid and the Earth's Unified Field all around us. Cathie's first book, *Harmonic 33*, was published in 1968 when he was a commercial pilot in New Zealand. Since then, Captain Bruce Cathie has been the premier investigator into the amazing potential of the infinite energy that surrounds our planet every microsecond. Cathie investigates the Harmonics of Light and how the Energy Grid is created. In this amazing book are chapters on UFO Propulsion, Nikola Tesla, Unified Equations, the Mysterious Aerials, Pythagoras & the Grid, Nuclear Detonation and the Grid, Maps of the Ancients, an Australian Stonehenge examined, more.
255 PAGES. 6X9 TRADEPAPER. ILLUSTRATED. $15.95. CODE: TEG

THE BRIDGE TO INFINITY
Harmonic 371244
by Captain Bruce Cathie
Cathie has popularized the concept that the earth is crisscrossed by an electromagnetic grid system that can be used for anti-gravity, free energy, levitation and more. The book includes a new analysis of the harmonic nature of reality, acoustic levitation, pyramid power, harmonic receiver towers and UFO propulsion. It concludes that today's scientists have at their command a fantastic store of knowledge with which to advance the welfare of the human race.
204 PAGES. 6X9 TRADEPAPER. ILLUSTRATED. $14.95. CODE: BTF

THE HARMONIC CONQUEST OF SPACE
by Captain Bruce Cathie
Chapters include: Mathematics of the World Grid; the Harmonics of Hiroshima and Nagasaki; Harmonic Transmission and Receiving; the Link Between Human Brain Waves; the Cavity Resonance between the Earth; the Ionosphere and Gravity; Edgar Cayce—the Harmonics of the Subconscious; Stonehenge; the Harmonics of the Moon; the Pyramids of Mars; Nikola Tesla's Electric Car; the Robert Adams Pulsed Electric Motor Generator; Harmonic Clues to the Unified Field; and more. Also included are tables showing the harmonic relations between the earth's magnetic field, the speed of light, and anti-gravity/ gravity acceleration at different points on the earth's surface. New chapters in this edition on the giant stone spheres of Costa Rica, Atomic Tests and Volcanic Activity, and a chapter on Ayers Rock analysed with Stone Mountain, Georgia.
248 PAGES. 6X9. PAPERBACK. ILLUSTRATED. BIBLIOGRAPHY. $16.95. CODE: HCS

ELECTROGRAVITICS SYSTEMS
Reports on a New Propulsion Methodology
edited by Thomas Valone
An anthology of two rare, unearthed reports of the secret work of T. Townsend Brown. The first report, *Electrogravitics Systems*, was classified until recently, and the second report, *The Gravitics Situation*, is a fascinating update on Brown's anti-gravity experiments in the early 50s. Also included, Dr. Paul LaViolette's research paper on the B-2 as a modern-day version of an eletrogravitics aircraft—a literal U.S. anti-gravity squadron!
116 PAGES. 6x9 PAPERBACK. ILLUSTRATED. $15.00. CODE: EGS

THE HOMOPOLAR HANDBOOK
A Definitive Guide to Faraday Disk & N-Machine Technologies
by Thomas Valone, M.A., P.E.
The second book from Tom Valone, author of *Electrogravitics Systems* and well-known free energy/anti-gravity scientist, is a milestone work on permanent magnet free energy devices. This book is packed with technical information with chapters on the Faraday Disc Dynamo, Unipolar Induction, the "Field Rotation Paradox," the Stelle Homopolar Machine, the Trombly-Khan Closed-Path Homopolar Generator, the Sunburst Machine, Experimental Results with Various Devices, more.
180 PAGES. 6x9 PAPERBACK. ILLUSTRATED. REFERENCES, APPENDIX & INDEX. $20.00. CODE: HPH

UNDERGROUND BASES & TUNNELS
What is the Government Trying to Hide?
by Richard Sauder, Ph.D.
Working from government documents and corporate records, Sauder has compiled an impressive book that digs below the surface of the military's super-secret underground! Go behind the scenes into little-known corners of the public record and discover how corporate America has worked hand-in-glove with the Pentagon for decades, dreaming about, planning, and actually constructing, secret underground bases. This book includes chapters on the locations of the bases, the tunneling technology, various military designs for underground bases, nuclear testing & underground bases, abductions, needles & implants, military involvement in "alien" cattle mutilations, more. 50 page photo & map insert.
201 PAGES. 6x9 PAPERBACK. ILLUSTRATED. $15.95. CODE: UGB

UNDERWATER & UNDERGROUND BASES
Surprising Facts the Government Does Not Want You to Know
by Richard Sauder
Dr. Richard Sauder's brand new book *Underwater and Underground Bases* is an explosive, eye-opening sequel to his best-selling, *Underground Bases and Tunnels: What is the Government Trying to Hide?* Dr. Sauder lays out the amazing evidence and government paper trail for the construction of huge, manned bases offsore, in mid-ocean, and deep beneath the sea floor! Bases big enough to secretly dock submarines! Official United States Navy documents, and other hard evidence, raise many questions about what really lies 20,000 leagues beneath the sea. Many UFOs have been seen coming and going from the world's oceans, seas and lakes, implying the existence of secret underwater bases. Hold on to your hats: Jules Verne may not have been so far from the truth, after all! Dr. Sauder also adds to his incredible database of underground bases onshore. New, breakthrough material reveals the existence of additional clandestine underground facilities as well as the surprising location of one of the CIA's own underground bases. Plus, new information on tunneling and cutting-edge, high speed rail magnetic-levitation (MagLev) technology. There are many rumors of secret, underground tunnels with MagLev trains hurtling through them. Is there truth behind the rumors? *Underwater and Underground Bases* carefully examines the evidence and comes to a thought provoking conclusion!
264 PAGES. 6x9 PAPERBACK. ILLUSTRATED. BIBLIOGRAPHY. INDEX. $16.95. CODE: UUB

THE TIME TRAVEL HANDBOOK
A Manual of Practical Teleportation & Time Travel
edited by David Hatcher Childress
In the tradition of *The Anti-Gravity Handbook* and *The Free-Energy Device Handbook*, science and UFO author David Hatcher Childress takes us into the weird world of time travel and teleportation. Not just a whacked-out look at science fiction, this book is an authoritative chronicling of real-life time travel experiments, teleportation devices and more. *The Time Travel Handbook* takes the reader beyond the government experiments and deep into the uncharted territory of early time travellers such as Nikola Tesla and Guglielmo Marconi and their alleged time travel experiments, as well as the Wilson Brothers of EMI and their connection to the Philadelphia Experiment—the U.S. Navy's forays into invisibility, time travel, and teleportation. Childress looks into the claims of time travelling individuals, and investigates the unusual claim that the pyramids on Mars were built in the future and sent back in time. A highly visual, large format book, with patents, photos and schematics. Be the first on your block to build your own time travel device!
316 PAGES. 7x10 PAPERBACK. ILLUSTRATED. $16.95. CODE: TTH

THE SEARCH FOR A NEW ENERGY SOURCE
edited by Dr. Gary L. Johnson
Johnson examines sacred texts and modern physics and constructs a free energy device. Material on unexplained phenomena, ball lightning, tornadoes, the earth's magnetic field, dowsing, UFOs, and gravitational anomalies. Also discussed are Tesla, Moray, Newman, Bearden and others. Chapters on Vapor Canopy; Ice Shell; Interplanetary Ice; Heat Balance; Fire from Heaven; Current Loops in the Earth's Core; Field, Aether, or Action-at-a-Distance; Patents; more.
263 PAGES. 6x9 PAPERBACK. ILLUSTRATED. $20.00. CODE: SNES

TECHNOLOGY OF THE GODS
The Incredible Sciences of the Ancients
by David Hatcher Childress

Popular *Lost Cities* author David Hatcher Childress takes us into the amazing world of ancient technology, from computers in antiquity to the "flying machines of the gods." Childress looks at the technology that was allegedly used in Atlantis and the theory that the Great Pyramid of Egypt was originally a gigantic power station. He examines tales of ancient flight and the technology that it involved; how the ancients used electricity; megalithic building techniques; the use of crystal lenses and the fire from the gods; evidence of various high tech weapons in the past, including atomic weapons; ancient metallurgy and heavy machinery; the role of modern inventors such as Nikola Tesla in bringing ancient technology back into modern use; impossible artifacts; and more.
356 PAGES. 6x9 PAPERBACK. ILLUSTRATED. BIBLIOGRAPHY. $16.95. CODE: TGOD

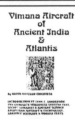

VIMANA AIRCRAFT OF ANCIENT INDIA & ATLANTIS
by David Hatcher Childress, introduction by Ivan T. Sanderson

Did the ancients have the technology of flight? In this incredible volume on ancient India, authentic Indian texts such as the *Ramayana* and the *Mahabharata* are used to prove that ancient aircraft were in use more than four thousand years ago. Included in this book is the entire Fourth Century BC manuscript *Vimaanika Shastra* by the ancient author Maharishi Bharadwaaja, translated into English by the Mysore Sanskrit professor G.R. Josyer. Also included are chapters on Atlantean technology, the incredible Rama Empire of India and the devastating wars that destroyed it. Also an entire chapter on mercury vortex propulsion and mercury gyros, the power source described in the ancient Indian texts. Not to be missed by those interested in ancient civilizations or the UFO enigma.
334 PAGES. 6x9 PAPERBACK. ILLUSTRATED. $15.95. CODE: VAA

LOST CONTINENTS & THE HOLLOW EARTH
I Remember Lemuria and the Shaver Mystery
by David Hatcher Childress & Richard Shaver

Lost Continents & the Hollow Earth is Childress' thorough examination of the early hollow earth stories of Richard Shaver and the fascination that fringe fantasy subjects such as lost continents and the hollow earth have had for the American public. Shaver's rare 1948 book *I Remember Lemuria* is reprinted in its entirety, and the book is packed with illustrations from Ray Palmer's *Amazing Stories* magazine of the 1940s. Palmer and Shaver told of tunnels running through the earth—tunnels inhabited by the Deros and Teros, humanoids from an ancient spacefaring race that had inhabited the earth, eventually going underground, hundreds of thousands of years ago. Childress discusses the famous hollow earth books and delves deep into whatever reality may be behind the stories of tunnels in the earth. Operation High Jump to Antarctica in 1947 and Admiral Byrd's bizarre statements, tunnel systems in South America and Tibet, the underground world of Agartha, the belief of UFOs coming from the South Pole, more.
344 PAGES. 6x9 PAPERBACK. ILLUSTRATED. $16.95. CODE: LCHE

ATLANTIS & THE POWER SYSTEM OF THE GODS
Mercury Vortex Generators & the Power System of Atlantis
by David Hatcher Childress and Bill Clendenon

Atlantis and the Power System of the Gods starts with a reprinting of the rare 1990 book *Mercury: UFO Messenger of the Gods* by Bill Clendenon. Clendenon takes on an unusual voyage into the world of ancient flying vehicles, strange personal UFO sightings, a meeting with a "Man In Black" and then to a centuries-old library in India where he got his ideas for the diagrams of mercury vortex engines. The second part of the book is Childress' fascinating analysis of Nikola Tesla's broadcast system in light of Edgar Cayce's "Terrible Crystal" and the obelisks of ancient Egypt and Ethiopia. Includes: Atlantis and its crystal power towers that broadcast energy; how these incredible power stations may still exist today; inventor Nikola Tesla's nearly identical system of power transmission; Mercury Proton Gyros and mercury vortex propulsion; more. Richly illustrated, and packed with evidence that Atlantis not only existed—it had a world-wide energy system more sophisticated than ours today.
246 PAGES. 6x9 PAPERBACK. ILLUSTRATED. $15.95. CODE: APSG

A HITCHHIKER'S GUIDE TO ARMAGEDDON
by David Hatcher Childress

With wit and humor, popular Lost Cities author David Hatcher Childress takes us around the world and back in his trippy finalé to the Lost Cities series. He's off on an adventure in search of the apocalypse and end times. Childress hits the road from the fortress of Megiddo, the legendary citadel in northern Israel where Armageddon is prophesied to start. Hitchhiking around the world, Childress takes us from one adventure to another, to ancient cities in the deserts and the legends of worlds before our own. Childress muses on the rise and fall of civilizations, and the forces that have shaped mankind over the millennia, including wars, invasions and cataclysms. He discusses the ancient Armageddons of the past, and chronicles recent Middle East developments and their ominous undertones. In the meantime, he becomes a cargo cult god on a remote island off New Guinea, gets dragged into the Kennedy Assassination by one of the "conspirators," investigates a strange power operating out of the Altai Mountains of Mongolia, and discovers how the Knights Templar and their off-shoots have driven the world toward an epic battle centered around Jerusalem and the Middle East.
320 PAGES. 6x9 PAPERBACK. ILLUSTRATED. BIBLIOGRAPHY. INDEX. $16.95. CODE: HGA

One Adventure Place
P.O. Box 74
Kempton, Illinois 60946
United States of America
Tel.: 815-253-6390 • Fax: 815-253-6300
Email: auphq@frontiernet.net
http://www.adventuresunlimitedpress.com
or www.adventuresunlimited.nl

ORDERING INSTRUCTIONS

✓ Remit by USD$ Check, Money Order or Credit Card

✓ Visa, Master Card, Discover & AmEx Accepted

✓ Prices May Change Without Notice

✓ 10% Discount for 3 or more Items

SHIPPING CHARGES

United States

✓ Postal Book Rate { $3.00 First Item
50¢ Each Additional Item

✓ Priority Mail { $4.00 First Item
$2.00 Each Additional Item

✓ UPS { $5.00 First Item
$1.50 Each Additional Item

NOTE: UPS Delivery Available to Mainland USA Only

Canada

✓ Postal Book Rate { $6.00 First Item
$2.00 Each Additional Item

✓ Postal Air Mail { $8.00 First Item
$2.50 Each Additional Item

✓ Personal Checks or Bank Drafts MUST BE

USD$ and Drawn on a US Bank
✓ Canadian Postal Money Orders OK

✓ Payment MUST BE USD$

All Other Countries

✓ Surface Delivery { $10.00 First Item
$4.00 Each Additional Item

✓ Postal Air Mail { $14.00 First Item
$5.00 Each Additional Item

✓ Payment MUST BE USD$

✓ Checks and Money Orders MUST BE USD$
and Drawn on a US Bank or branch.

✓ Add $5.00 for Air Mail Subscription to
Future *Adventures Unlimited* Catalogs

SPECIAL NOTES

✓ RETAILERS: Standard Discounts Available

✓ BACKORDERS: We Backorder all Out-of-

Stock Items Unless Otherwise Requested

✓ PRO FORMA INVOICES: Available on Request

✓ VIDEOS: NTSC Mode Only. Replacement only.

✓ For PAL mode videos contact our other offices:

Please check: ☑

☐ This is my first order   ☐ I have ordered before

Name

Address

City

State/Province   Postal Code

Country

Phone day   Evening

Fax

Item Code	Item Description	Qty	Total

Please check: ☑

Subtotal ➡

Less Discount-10% for 3 or more items ➡

☐ Postal-Surface   Balance ➡

☐ Postal-Air Mail   Illinois Residents 6.25% Sales Tax ➡
(Priority in USA)   Previous Credit ➡

☐ UPS   Shipping ➡

(Mainland USA only) Total (check/MO in USD$ only) ➡

☐ Visa/MasterCard/Discover/Amex

Card Number

Expiration Date

10% Discount When You Order 3 or More Items!

224